THE PREMIER GUIDEBOOK FOR BUSINESS GLOBE TROTTERS

eBiZ
guides

LIBYA

www.eBizguides.com

CREDITS

It is always good to know who worked hard to produce this guide. Many wonderful stories and people are present, here is the list:

PRODUCTION

Producer: Pascal Belda

Regional Project Direction: Frederic Van de Vyver

Local Project Director: Elisa López Moriarty

Local Guide Developer: Jaume Sugrañes

Research and Market Development: Nicholas Bruneau

Advertising Agency: Maya Studios

EDITION

Written by: eBizguides

with the collaboration of The General People's Committee For Tourism

Information provided by the G.P.C. for Tourism and Mohamed Ghattour & Co.

Edited by: Jaume Sugrañes and Tim Koop

Art Direction: Elisa L. Fuentes

Pictures: Ahmed Tarhuni, Elisa López Moriarty and Jaume Sugrañes

ACKNOWLEDGEMENTS

We would like to thank the following people for their collaboration and support:

H.E. Saif Al Islam Ghaddafi, Chairman of the Ghaddafi Charity Foundation. H.E. Ammar El Tayeff, Secretary of the G.P.C. for Tourism – H.E. Don Francisco Tapia, former Ambassador of Spain in Libya – H.E. Jean Jacques Beaussou, former Ambassador of France in Libya - H.E. Dr. Aribi A. Mazouz, Under Secretary of the G.P.C. for Tourism – H.E. Ali M. El Sherif, Under Secretary of the G.P.C. for Economy and Trade – Mr. José María Pascual, Advisor of the Spanish Embassy in Libya – Mr. Ali E. A. Chadouri, Chairman of the Department of Antiquities – Mr. Monder Ramadan, General Manager of the Ghaddafi Charity Foundation – Mohamed H. Layas, Chairman of the Libyan Arab Foreign Bank – Mr. Zaroug, Chairman of the Libyan Arab Foreign Investment Company – Mr. Dabnon, Chairman of the Libyan Arab Airlines – Eng. Abdalah, Chairman of the Libyan Business Council – Eng. Sulieman Abu sa'a, Chairman of Aman Company for Tires – Dr Ties Tiessen, General Manager of Wintershall – Mr. Musa El-Naas, General Manager of Al-Madar Telecom – Mr. Fadel Fellah, General Manager of Schlumberger – Mr. Joseph Pisani, Mr. Mark Gauci & Mr. Salvinu Farrell from Corinthia Hotels International – Capt. Sabri S. Shadi & Mr. Said Albarouni from Afriqiyah – Mr. Mohamed Oun & Mr. Ahmed Tarhuni from Zueitina – Eng. Nasardeen Elhawat, General Manager of Trucks & Buses Co. – Mr. Graciano Rodríguez, Chairman of Repsol – Mr. Mohamed Ghattour & Mr. Muuftah Elklaiby from Mohamed Ghattour & Co. – Mr. Ali Siala – Mr. Osama Ghattour & Mr. Joseph De Bono from Wadizzir Investments– Mr. Abdurrazag Gherwash, Chairman of Winzrik – Mr. Rajab Shiglabu – Mr. Mohamed Suleimanm, General Manager of Libyaninvestment.com – Mr. Fituri M. Saied, General Manager of Libyan Travel & Tourism Company – Eng. Abdulrahim N. Al-Muntasser, Chairman of the National Development & Real Estate Investment Co. – Dr. Salem M. Bengharbia, Chairman of the Libyan Insurance Co. - Mr. Godwin Pullicino, Country Manager of British Airways – Mr.Joerg Pigaht, Country Manager of Petro-Canada – Tor Bjormulf Cund, General Manager of Hydro.

We would like to specially thank the following collaborators & friends in Libya for their professional assistance and kindness during our stay: Eng. Abdurrazag Abulgasim, Mr. Ahmed Aziz, Eng. Sulieman S. Youssef, Eng. Hussein Krawa, Mr. Adel Farina, Mr. Omar Said, Mustafa, Mrs. Huda M. Ghith, Col. Ashur Emgeg, Mohamed Zeglam, Mohamed Taher, Mrs. Siala, Fatima, Ranko Grudic, Gradimic Dragic, Tony Tambone, Jamila Durran, Haithem M. Elkeeb, Ashraf Elmabrouk, Ali M. H. El Musrati, all the team from the Tourism Investment Board and all the other good friends that we will never forget.

We would also like to thank the chairmen of the following companies for the sponsorship and support and without whom this eBizguide would not have been possible:

General People's Committee for Tourism – Libyan Arab Foreign Investment Company – Libyan Arab Foreign Bank – Zueitina – Trucks & Buses Company – Schlumberger – Alcatel – Aman Company for Tires – Corinthia Hotels International – Afriqiyah Airways – Engineering Industries Company – Winzrik – Libyan Youth & Students Travel Company – World Investment News Ltd.

World Investment New Ltd. - Fourth Dollard House, Wellington Quay

Dublin 2 - Ireland

Customer Service Tel: + 34 91 345 66 25/ Fax: +34 91 787 38 89

E-mail: info@ebizguides.com

www.ebizguides.com

Printed by: IM. Grafihumo

Printed in Spain - EU

Legal Deposit: TO-478-05

ISBN 84-933978-2-2

How to use this guide?

The goal of eBizguides is to offer extensive economic and investment information on a country, with a focus on the Top Companies, whilst offering the best tourism and leisure information for your spare time.

With this, we believe that our readers can get fully acquainted with the country, before investing on a long-term basis.

This book is comprised of three major sections:

General Information

Background information on the country is coupled with the main business resources available to you.

The Country Economy

This section of the book starts off with an in-depth look at the Investment and Legal Framework, as well as the Country Economy itself.

Through personal interviews with the most important personalities in the country, we then provide our readers with privileged information and sector analysis, including introductions written by key decision makers in the sector, such as Government Ministers.

You will also find Fact Files of all the major corporations, outlining their business activities, key projects and any investment opportunities.

We consider this to be some of the most invaluable information for Business Globetrotters.

Leisure

For you to fall in love with the country, we also showcase the best spots to visit, essential areas to discover, the top hotels and restaurants, and much more.

Icons

eBiz Recommended

These companies were particular favourites of our team during their stay, this stamp is mostly given to restaurants, hotels, etc.

eBiz Recommended Partner

Our team has been interviewing the heads of many companies, and when they choose to allocate this stamp, it is because they themselves have received a reliable service from the company

Notice

This indicates that the information is important and something to take notice of when planning, or during, your stay.

Legends

Proverbs

These local phrases will help you to better understand the popular culture in the country. Perhaps you can even use them yourself.

Top Companies

This is a listing of the Top Companies that the team encountered during their time in the country.

Blue Notes

This is useful insider information that will help you to assess the country, do business and enjoy your stay.

Libyan Arab
Foreign Bank

Your **Best**
Partner
Linking **Libya**
to the
World

Libyan Arab Foreing In

From humble beginnings in the mid-seventies, LAFIC
continent in more than 25 countries and in diverse s
mining and telecommunications, never wavering in its

Our Goals

To generate revenues for the sharehold

To promote inter-African trade.

To invest in economically viable projects.

To diversify the inv

To practice a sustainable utilization of the environment.

To utilize the local

YOUR GATEWAY T

Tel: +218 21 4890 / 91

+ 218 21 47814

Gharian - L

estment Company

usiness interests now span the four corners of the African
rs from hotels & real estate, industry ,agriculture , trade to
vention in Africa's potential for growth and development.

Exchange of technology.

To integrate the African countries economies, regionally and globally

Our Policies

nents sector wise and geographically.

naterials in a profitable and professional manner.

o INVEST IN LIBYA

4771790   Fax: + 218 21 3600706 – 3614892

ya

Index

Did You Know?

* Tripoli (Tri-polis) means "the three cities", in reference to Sabratha, Leptis Magna and Oea.

* The flag of Libya is a green rectangle.

* Libya is the only country in the world where boxing is not allowed as it is considered too violent.

* Like in many other Muslim countries, Libyans put their hand on their heart after shaking hands with someone; it is a sign of friendship and respect.

* Libya has five sites declared as world heritage sites by the UNESCO; these are Ghadames, the prehistoric paintings of Akakus, Leptis Magna, and Sabratha & Cyrene.

* Libya is the fourth biggest country on the African continent after Sudan, Algeria and Nigeria.

* Libya is the only country in the world that is not irrigated by a permanent river (but it has a man made river).

Libya Fact File

Area:	1,774,440 sq km
Population:	4.389.739 (last census in 1995)
	note: this figure is now more in the region of 6.1 million
Population growth rate:	(approximately)
Main Cities:	2.2%
	Tripoli, Benghazi, Sirt, Sebha, Misurata, El Beda, Zawia, El
International Airports:	Khoms, Tobruk and Derna.
International Ports:	Tripoli Airport, Benghazi, and Khoms
Climate:	Tripoli, Benghazi and Sabha
	Mediterranean in the north. Dry, extreme desert in the
Electricity:	interior
International Telephone code:	220 Volts
Language:	218
Fiscal Year:	Official language Arabic, second language English
Time:	Calendar year
Currency:	GMT +2 hours
	Libyan Dinar (LD)

GENERAL INFORMATION

" GADHAFI PRIZE FOR HUMAN RIGHTS "

INTRODUCTION BY H.E. ENG SAIF AL-ISLAM GHADDAFI

It is a pleasure for me to introduce the first business guide on Libya for the eBizguides Pan-African initiative. This guide will provide those visiting the country with business and tourist tips and allow them to operate efficiently in the country.

From the Eastern to the Western Region where the country embraces the Mediterranean with over two thousand kilometers of incredible sea shores, or from the Central to the Southern region where the breathtaking Great Desert lies, visitors will be amazed by all the country's natural treasures.

The investment and business potential represent yet another treasure of the country. The recent opening onto the international market makes it easier to collaborate, trade, and invest in Libya. Moreover, Libya is among the most stable and safe countries of the world. It is one of the sought after destinations for international investors.

When considering Libya as an investment or tourist destination, it is important to consider the many advantages the country has to offer. Libya has a unique strategic geographical position: it is the gateway to Africa. The country has a young and well educated population that can respond to the human resources needs of the international business community. The oil and gas sector has and will continue, to offer many opportunities for investment. The industrial sector benefits from a low cost of raw materials and energy. The tourism sector is mainly underdeveloped and offers a fertile ground for investment.

Further more, substantial efforts have been recently undertaken in order to modernize the Libyan economy and the Libyan society. Underlying these efforts are reforms that will attract foreign investment and raise the standard of living. Libya is indeed collaborating with many international organizations such as the World Bank and the International Monetary Fund. Other States, Governments, and international experts have joined in to help Libya revamp its economy.

In order to explore all of the above opportunities, Libya extends its invitation to governments, investors and international businessmen to get to know more about the country and to participate in its economic development.

"H.E. Eng. Saïf Al Islam Ghaddafi"

INTRODUCTION BY H.E. AMMAR ELTAIF SECRETARY OF THE G.P.C. FOR TOURISM

When eBizguides asked me to write a presentation for this impressive work, I did not hesitate as I found it a great opportunity to celebrate my love of Libya at one time and to glory in it at another...... and to write more and more about its charms... Why not?... It is it my beloved country! ... And who doesn't love his country!!!!

Libya is present in Homer's epics, in the writings of Herodotus... and in the tales of sailors in the Mediterranean while they were cruising it east to west, north to south, leaning on its shores when the waves irritated them and sleeping in its harbours when exhausted from travel.

Libya has always been present in caravan traders' songs and melodies as they crossed the great desert between the Mediterranean and Central Africa. They smelled strongly of spiced perfumes and were adorned with ostrich feathers and ivory.....

The oases of Libya in Murzak, Jalo and Sebha, are places where the traveller can rest and quench his thirst...

And the cities of Ghadames, Ghat and Murzak tell of old civilizations and are centres for culture and teaching...

It is Libya where the civilization from more than 5000 years ago can still be seen engraved on the rocks of the Akakus Mountains in Metkhandush, as well as in Tadrarat and Emsak. The Wades and Germant civilizations still decorate Zankikra Mountain which rests in old Germa. From there, the Ubari sands fade away and Gabar Aoun Lake suddenly appears in front of you as you take in and dwell upon what you see all around you. Tale

GENERAL INFORMATION ABOUT LIBYA

after tale are told by the whispers of the high mountains down to the deep wades, twirling through the palm trees before slipping into deep silence and disappearing across the desert. where sometimes diminishes within Ubari sands where suddenly Gabar Aoun lake appears in front of you where you dewela with your vision in all directions to hear a tale after tale told by the high mountains to deep wades .... to hug the palm trees , then it slips in deep silence across a hugs desert.

Then, with your hands, you touch Ghadames, the desert nymph, and find yourself wondering in her unique streets where the sights will take you back to centuries ago as you listen to its tales from an elderly man still sitting in front of his dwelling....

Now you would find yourself intrigued in front of old Ghat with its castle and streets in old Ghat and Tonin.... Then from Alfut, leaning on the Akakus Mountains to the west, you hurry east toward Wawo Enamous and eastern Awinat with its thrilling scenes and old engravings..... Here you oversee the Jaboub oasis, and Bzima and Kufra... On your way to the north you will certainly pass by Zwilla to get a glimpse of Islamic civilization...

If you get to the southern shores of the Mediterranean in the east of Libya where the pentapolis (Shahat, Cyrine, Sousa, Tolmitha, Tucera and Benghazi) in the Green Mountain is, you will see the Greek civilization from the first century B.C. You will hear about the civilization from the three graceful ladies in Cyrene ....by Sousa Harbour and by the Tolmitha columns. Your sight seeing will be completed on its splendid beaches in Ras-Elhilal and Alhamama...

If you continued in your way toward west alongside beaches, you will certainly stop over at Gdabya with its Islamic Fatmi Palace, then to Sultan nearby old Sirt and its Phoenician civilization from thousands of years ago...... from there, your curiosity will attract you to Liptus Magna, with its Phoenician - Roman civilization, to swim in the famous Hadrian Baths and to shop in the Punic Market and explore its theatres and streets...

And if you enter (Oea) Tripoli - Modern Tripoli with its old, historical city - you will be told tales of ancient times and of its famous castle where the Romans, Vandals, Knights of Malta and American fleet were all broken and where the Philadelphia Mast still stands as a testament to all that...

There is also Sabratha in the west of Libya which tells another story of another era in history from the Phoenician to Roman civilizations with its famous theatre cantered in a city which tells tales of one thousand nights, embracing the Mediterranean waves in the middle of olive tree farmlands.
Do I tell the traveller tourist of the eighth wonder in the modern world?

It is the great man-made river which was engineered by a man in Libya and which runs for more than five thousand kilometres, carrying millions of cubic meters of water from the middle of the desert from the south of Libya to its north in the biggest developmental project in modern history.

Above all else, you'll meet a hospitable and sincere people aspiring for freedom, peace and progress.

Ammar Elmabrouk Eltaif
Secretary of the General People's
Committee for Tourism

HISTORY

The roots of man settling in Libya go back to the prehistoric period or the Stone Age, starting around 10,000 B.C. to 2000 B.C. This period left evidence of human settlements in Libya through a wealth of paintings and engravings left on stones inside mountain caves that are generally located in the southern part of Libya. These show us the way they lived as well as animals and living creatures and green land which has since then changed into desert areas. Groups of people or Libyan tribes lived around 10th to 11th century BC in this area and they had politica and commercial ties with the ancient Egyptians which resulted in big Libyan influences in Egyptian society, and they left a rich heritage giving us a taste for their way of life. Due to its strategic location, Libya has always seen invaders, such as the Phoenicians who were attracted to the strategic position of the North African coastline. They established settlements during the 15th century BC and their cities during 8th century BC at Sabratha, Tripoli (Oae) and Leptis Magna and near to Sirt (Macomadis); and Charax City as well. Little visible evidence of their settlements exists today. As for the Greeks who lived in Cyrenica, the eastern part of Libya, during the 7th century BC, and after, they built their cities Cyrene (Shahhat), Tocharia (Tokra) Appolonia (Sousa) the Cyrene Port, Ptlemos (Tulmitha), Barce (El-Marj) and Bernice (Bengahzi). The Phoenicians, in the western part of Libya, established their commercial centres in the 15th and 14th century BC which later became great cities in the 8th century BC such as Oea (Tripoli), Sabratha, Leptis Magna, Sirt and others. The Roman influence spread along the Libyan coast after the Romans took control of the Mediterranean Sea. Many significant cities, Phoenician and Greek, were used by Romans as ports for trading slaves, ivory, precious metals, olive oil and animals across the desert. Such trade became prosperous after the fall of Ghedames (Sedamos) in the year 19 BC at the hands of the Romans. In the latter part of 4th century AD, the Libyan coastal area became Christian in combination with other religious minorities also present. This situation led to many destructive divisions in Libya. After this period of time, Libya was invaded and then occupied by the Windalis. Islam spread in Libya at the beginning of the 7th century AD following the Islamic victory and conquest of North Africa. Libya suffered many conquests and occupations by European powers such as the Sicilians, Spanish and Maltese (Saint John Knights). Then Libya was under the control of the Ottoman Empire. A series of semi-independent Turkish rulers governed Libya from 1711 to 1835 AD Life in Libya was, in general, relatively quite until the Italian conquest in the year 1911. After that, many events took place leading up to the British control of the Tripoli province and the French control of the Fazan area. This was the case until the country was granted a semi-independent state by the United Nations in 1952.

Many old Libyan paintings are incredibly well conserved"

As a result of the economic and political decline in Libya and the general heading towards the Arabic nationalist ideology prominent in the Arab countries, the Revolution in Libya took place on the 1st of September 1969 led by Colonel *Moamer Al-Gaddafy*. Since then, Libya has started a new era characterized by significant accomplishments in different political, economic and social fields in large.

The impressive roman city of Sabratha

A Brief History of Libya and its Prominent Relics

Libya, with its vast two millions square kilometres, is characterized by its rich archaeological heritage rarely found in other areas in the world. Prehistoric relics can be found on most of this vast area or immersed in the seawater along the coast. Such significant heritage was well preserved over the last ten thousand years due to the dry weather conditions that prevailed over that time. The period from which of the oldest discovered tools that belonged to the stone culture, is referred to as the Lower Stone Age, which took place at the beginning of the Pleistocene epoch 2-5 million years ago. Such tools were just smooth stones shaped at one side to make a sharp edge. The technique and use of such tools improved with the passing of time and were developed in accordance with greater human knowledge and experience. The simple stone culture developed into the Achoulian culture, which was famous for its hand-axes

shaped as pears. In its late stages such Achoulian axes reached a high level of quality as its shapes and lengths were diversified. They were long, sharp-ended, and almond, pear or heart shaped. The sharpening of the axe later covered all the body of the axe so that it had a beautiful shape. This technique was improved until it reached such a high standard that it became an artistic masterpiece. The makers or these axes kept them for themselves and did not use them as tools, as was proven by the studies that were carried out on the use of stone tools. Those two cultures covered all the Lower Stone Age, which is considered the longest Stone Age. Due to the desertification of this area, the serial layers of the archaeological levels were completely destroyed due to erosion factors at most of the desert archaeological sites. Therefore, most of the discovered stone tools were only surface collections and were not the result of programmed ground layer excavations. This makes the comparative studies for such tools, with other discovered tools discovered through layer sequence excava-

21

tions a basic reference for the determination of the time period of such surface discovered tools. The geomorphologic studies and the ancient climate facilitate the reference of such tools to their right time and environmental stage.

The Middle Stone Age started with the rise of the Mesolithic culture, which achieved specific technical advances. After using the sharpened stones, the user started to use the fragments of the stones to make his tool instead of the stone itself. The Ataryan culture was a branch of the Mesolithic culture in North Africa and it was distinguished by the use of blades in its tools.

This Ataryan culture invented blades and was the pioneer in using them. In addition to the stone tools belonging to different ages throughout Libya, there are hundreds of thousands of artistic drawings (paintings) coloured and engraved on the stone cliffs at the valleys and at some specified sites. Some of these can be found at Acacus Mountains, Missak Hill, Hamada Hamra High Land, Owenat Mountain, Arkno Mountain and in Tebisti Mountain. The significance of such drawings and paintings belongs to different stages of history, and can be specified by the subject of the painting or the method by which it was engraved. The subject of these paintings include the drawings of elephants, giraffes, alligators, rhinoceroses, buffalos, and other types of extinct cows as well as lions, bears, leopards, deer, wild cows, birds, ostriches, and finally the engravings of domesticated horses, camels and drawings of carts that are drawn by four horses. The other important source of Libyan history is what the father of history (Herodotus) wrote in the 4th century B.C. when he visited Cyrene City and met with many Libyan personages

The Greek temple of Zeus at Cyrene founded in 631 B.C.

who had extensive knowledge in the region and its people. From them he got his information concerning the number of the Libyan tribes including the Germent tribes i.e. the Germenions.

The Germain (Germents):

In his 4th book, Herodotus reported that the natives of Germa pursued land cultivation, horse breeding and used chariots (carts pulled by four horses), and that their cattle had big horns. Archaeological research has shown that the influence of the Germent culture extended to the north to include parts of Alhamada Alhamra (the red highland) and south to Barjuj Wadi, Emsak highland. This is apparent from the form of tombs, which were mostly of the circular type and in the form of hands, which also have sacrifice tables carved in rocks. Many of such tombs are found along Wadi Alhyat (Alajal). The Germain were also masters in conveying underground water using underground canals from far away. They called these canals vertebrates. Some of these can still be seen nowadays in the south of Libya. The fortified village in Znkaker Mountain is one of the most ancient archaeological sites of the Germents. It was their capital before Germa was built. Germa archaeological relics were discovered beneath the Islamic Germa, which has an area of about 20 hectares. Some architectural landmarks bearing the Roman style and belonging to the 1st century AD were also discovered. However, the Roman connection was made far earlier than this as the Romans had waged a punitive campaign under the leadership of Balbous at the end of the 1st century BC. The Romans on the coast had a relationship with Germent who mastered and controlled the desert and its caravan routs until the Arab conquest when Germa was conquered by Aoqba Ben Nafeh Ahfehri (642 AD). Gasr Watwat near Germa, said to belong to the Roman period, has one of the Roman Funeral idols in the south, which resembles in its architecture style the Phoenician funeral idols in the north.

The magnificent 1st Century B.C. Theatre of Leptis Magna, the birthplace of Emperor Septimius Severus

Classic Ages

The Classical Ages, according to some historians, started first with the Phoenicians in the west in the 2nd half of the 2nd millennium BC then later with the Greek settlement in Green Mountain in the 7th century. The Greeks established the Pentapolis (The Five Cities) in Green Mountain, and the Phoenicians established the three cities: Oea, Leptis Magna and Sabratha. These cities were later given the name of Tripoli, which means the Three Cities.

The Greeks established their polis around 631 B.C. at a water spring, which they called Appolo Spring. The Greeks planned their rising city starting from Appolo Spring toward Batus Tomb, their first king, from there to the city square, "Agora," where public buildings and temples were scattered around. During the Roman era the city was given a Roman touch which was to make it rely on network planning, so streets were made perpendicular and to cross each other. As for beliefs and faith, both had also acquired a Roman touch through using Roman names instead of Greek ones. For example, Aphrodite became Venus and Athena became Manerva, Artimes into Diana and Bakus became Dunisos. Cyrene and the other Pentapolis participated in all sorts of creative activities and some of the region's natives and sons became famous as philosophers, poets and artists. The region also contributed much to Christian history such as Saman "Simon" the Cyrenic and the Saint Markus who was born and grew up near Shahhat before Roman idolaters dragged him into Alexandria in 70 AD.

Tripolis

The Tripolis was founded on the Libyan western coast among a set of commercial and trade stations established by Kenanits trade sailors who were mistakenly thought to be Phoenicians. The Kenanits mastered most of trade business in the Mediterranean Sea and played the role successfully between the different civilizations, peoples and continents in spite of differences in their cultures and races. They gained much influence and financial profits. Over time,

such trade stations became highly influential cities led by Carthage "Kariat Hadichat". This city was established by the Kenits themselves in 814 BC in order to get their hands on those newly established polis in the 6th century BC to avoid the Greek threat which was then starting to expand its influence west.

The Kenantis were successful in spreading their language and culture in North Africa and influenced it with their style and touch. They also spread their trade knowledge by starting camel caravans to the middle of Africa. In addition to this, they mastered irrigation and agriculture, especially wheat, barely, olive trees and palm trees. In the 1st century B.C, these polis fell to Roman influence after a long and bloody conflict. Roman influence and style could be seen in all aspects of life in the area, however, the Kenanit culture and language stood strong and remained in the area among the local residents until the time of the Arab con-

Mosaics from the Basilica of Justinian at the Sabratha Museum

quest in 642 AD. The ancient archaeological sites of the Tripolis, the Roman era and to a lesser extent the Byzantine era witnessed the biggest urban expansion during the 2nd century AD making it among the biggest Roman civilizations, as can be seen from the remains of the temples, theatres, play grounds with stairs, markets establishments and public areas all of which reflect the scale of progress and promotion seen by these polis. As a result the polis were given the title of municipality, then a colony in the 2nd century AD giving it the same status as the Roman Polis on the opposite coast of the Mediterranean. Its buildings had a great architectural artistic significance which consolidated the fundamentals of the genius Syrian artist, Polodorius, who was designated as the royal architectural engineer of the empire during Emperor Haderian's reign. Architecture of Hadrian bathes which were the biggest bathes in the Roman Empire outside Roma arose as another example of the urban and architectural renaissance, while the Basilica was an extremely beautiful architectural edifice.

Mosaic

The mosaics were flagstones, well decorated and engraved with fine colours bearing engineering or natural inscriptions to reflect the religion beliefs or dates of certain events. The small coloured stone pieces are compressed on a wet mud floor to fix it in its place and thus establish one of the most significant historical findings on which archaeologists can rely in their studies. The history of mosaic dates back to Mezopotamia. As for the mosaics discovered in Libya, they belong to the Alexandria School, which was famous for its precious and fine scenes. The mosaic used in Vila Abo-Umira in the city of Zletin in Libya is regarded as one of the most beautiful to have been found so far.

Cylene Villa (Dwelling)

The mosaic of Villa Cylene near the city of Khoms is a coun-

try palace for some wealthy people of Leptis Magna from the 2nd century, which was found to still be in good condition, almost complete with its water cisterns, baths, mosaic and coloured wall drawings. The mosaic represents scenes from the stadium in the city. This Villa was covered by sand after being deserted in the 5th century, due to internal disturbances, thus keeping it intact.

Islamic Architecture

A lot of archaeological evidence of past inhabitants in the area still remains along the coast and in the desert areas demonstrating the architectural styles of these people. Barca is considered one of the most important cities and it served as the first Islamic capital from the beginning of the

An old mosque among the beautiful ones of Tripoli

Islamic conquest; there is still much of evidence for its greatness. Both Ajdabiya and Sirt were the most important cities for conquest. They contain archaeological places of interest. Zuila was one of the most important cities in Sahara, which controlled a large desert empire and its white Mosques still show its greatness. Islamic Germa, Morzaq, Ghadames among others are still standing as important examples of what is called the local architecture. In the western part of Tripoli there stands a defence wall, which had been built over time by several governors of Tripoli starting with the Islamic conquest through to the Turkish period. Libya also has many landmarks and tombs such as the Tombs of Sahaba in Derna, the tomb Rwaifa Ben Thalit Alansari in AlBeida, the Mosque of Abdella Ebn Abi Sara in Augla and some remains of a fortress which was called Ribats. Some Mosques such as Abdulla Wahab Kaisi and An-Naga Mosque in Tripoli represent the type of architecture in Libya. Finally, the remaining Islamic archaeological cities are present in a lot of places with Sharuss City being one of them, in the area of the Western Mountain.

Tripoli Castle

As for Tripoli Castle, in spite of its long history, today what remains is not older than the 16th century AD. Influence of Spanish occupation and signs of Malta can still be clearly seen in spite of the Turkish and Italian additions. The Italian occupation preserved the urban planning executed by the second Ottoman reign at the start of the 20th century. However, the Italians erected several buildings bearing their characteristics and style, which was gothic and rococo. The Andalusian and Maghribi styles also had an affect on some of the buildings and constructions in Tripoli.

Underwater Archaeologies Objects

Libya (G.S.P.L.A.J) enjoys a long 1900 km coastline overlooking the southern shore bank of the Mediterranean Sea. The southern beaches of the Mediterranean are regarded as the richest places for underwater archaeologies, which luckily escaped the treasure hunters' exploitation. These long beaches and seashores still have in their depths many vessels and ships full of their shipments and loads which they were carrying when they sunk into the sea. These ships include Phoenician, Greek, Roman, Arabic and Turkish ships, and many more. The underwater archaeological sites also include several old buildings which flooded with seawater due to natural factors as can be seen at Greek Appolonia (Sousa) and Roman Leptis Magna ports.

THE COUNTRY TODAY

The last census in 1995 places the population at approximately 4.389.739 inhabitants. This figure is now more in the area of 6.1 million and some 700,000 foreigners. The annual growth rate is about 6%. Expatriates, working under government contracts to meet labour shortages, are largely citizens of other Muslim countries. Many technical and professional positions are filled by Eastern and Western Europeans. The total workforce is 1.357.063 million with 31% in industry, 27% in services, 24% in government and 18% in agriculture. Altogether, representatives of more than 100 nationalities live in Libya. The number of medical doctors and dentists reportedly increased nine-fold between 1970 and 1997, producing in the case of doctors a ratio of 1 per 732 citizens. The number of hospital beds tripled in the same period. Further progress has included the eradication of malaria and significant gains against trachoma, tuberculosis, and leprosy. In 2003, the infant mortality rate was 19.4 per 1000. Life expectancy for men is 71.2 years and for women 73.8 years. In the early 1980s, estimates of total literacy were between 50% and 60%, about 70% for men and 35% for women, but the gap has narrowed significantly due to increased female school attendance. The overall average in 2003 was approximately 87.4%.

View from the Corinthia Bab Africa Hotel of the gardens next to the Dat El Imad business complex

GOVERNMENT

Political System in the Great Jamahiriya

In the Great Socialist People's Libyan Arab Jamahiriya, the authority is in the hands of the Libyan people. This is pursued by the people through and in the basic conferences in which all citizens, men and women 18 and over, get together to discuss situations and make decisions.

Basic People's Conferences

These conferences directly pursue the authority, governance, control and management of the state's affairs. They issue legislation and decisions and are the only political reference. They make economic and social planning as well as public budgets. They choose and question and investigate their secretariats, executive and control committees, as well as the heads of people's courts. The people's basic conferences set the political relationship between the Great Jamahiriya and other countries. These conferences approve the treaties and agreements signed between the Great Jamahiriya and other counties and decide on peace and war. The people's basic conferences convene at least once every four months and it may be called for meetings in extraordinary sessions upon a call by the secretariat of the basic people's conference or by a request by the majority of the people's basic conference members or by a call from the secretariats of the General People's Congress. Every citizen is registered as a member in the basic people's conference in the area in which they live.

Secretariat of the Basic People's Conferences

The members of the conference directly choose the conference committee. This committee is concerned with drafting up the decisions of its conference and the execution of their follow up. It is also concerned with holding joint meetings with the conference people's committees, and coordinating with the secretariat conference people's of the shabya (province) committee, in which the people's basic conference is located, regarding the organizational questions. It also has the authority to investigate the basic conference people's committees, mainly with the conference secretary or any of its members, and forms investigation committees to that end. The basic people's conference chooses among its members, in a direct manner, the people's committees to manage the sectors. These committees specialize in the implementation of basic conference people's committees related to the competent sector. It is also in charge of managing production and service utilities of the basic conference people's committee, supervising the public utilities falling within the conference vicinity and exerting its authorities under the supervision of basic conference people committee.

The People's Committee of the Basic Popular Conference:

It is formed of the secretaries of People's Committees for sectors. The basic popular conference chooses its secretary and this committee shall be in charge of the following: Execution of the basic popular conference decisions within its area, execution of the General Peoples Committee decisions, the General People's Committees for Shabyat, General People's Committees for Sectors and the People's Committee of the Shabyia, preserving law and order within the conference's administrative area, supervising the follow up productive and services activities within its administrative range, controlling prices within the range of the basic popular conference. Setting programs concerned with creating employment opportunities for the conference member, reviewing applications regarding establishments of companies, changing their activities or merging them, issuing economic activities licenses for individuals within the conference area, providing public and administrative services for conference members and implementation of urban plannings. The basic popular conference enjoys official entity and independent financial status - each basic popular conference has an annual budget consisting of whatever was allocated to it from the general budget, local fees, revenues and taxes.

People's Conference of the Shabiat (Province)

The Great Jamahiriya is divided into 32 shabiats (provinces) which include 456 people's basic conference. The shabiat people's conference consists of basic people's conference secretariats within the geographic area of the shabiat and the secretaries of these basic conferences people's committees, syndicates, secretaries and unions and professional associations falling within its vicinity.

The People's Conference of the Shabiat is concerned with:

Choosing its secretaries, the secretary of the people's committees of shabiat and secretaries of people's committees for sectors. It is also concerned with choosing the secretary of planning, council of directors and the chairman as well as members of the people's court and the secretary of the people's committee for people's control apparatus. of the shabiat It is also concerned with drafting of basic people's conferences decisions at the shabiat level and transferring it to the General People's Conference, and to the competent people's committees for execution approval. Further, it is concerned with distributing the different budgets on the basic people's conferences and their sectors and investigating the chosen members and accepting their submitted resignations and relieving them from their duties.

The Sectors People's Committees of the Shabiat People's Conference:

Sectorial Peoples Committees are formed on the Shabiat level from secretaries of sector Peoples Committees of the basic popular conferences. The Shabiat Peoples Conference chooses the secretary of the Peoples Committee for each sector among those who were directly chosen by the people. The sector Peoples Committee of the Shabiat is concerned with the implementation of the basic popular conferences decisions inside the Shabiate as well as the execution and managing the public utilities and projects falling within the Shabiat range.

The People's Committee of the Sectors is Concerned with the following

Executing of the people's basic conferences decisions within the shabiat, also implementing and managing public utilities and projects falling within the shabiat geographic limits.

The Shabiat People's Committees

Consists of the secretaries of Peoples Committees of the basic popular conferences and sectors Peoples Committees secretaries of the Shabiat. The Shabiat Peoples Conference chooses its secretary among those who were directly chosen by the people, a secretariat for the Shabiat Peoples Committee is formed of its secretary and the sectors Peoples Committees secretaries and is concerned with: Implementation of the basic popular con-

View from the main mosque in Ghaddames

ference, decisions of the General Peoples Committee, the General Peoples Committee for Shabiats and the General Peoples Committees for sectors, preserving order and rule of law within the Shabiat range. Providing services and management of public utilities inside the Shabiat in coordination with the basic popular conferences Peoples Committees. Recommendation of transfer plans, transfer budget as well as the budget for running the sectors, concluding contracts for projects execution regarding transfer plans, establishment of investment projects, execution of the different programs for granting loans, supervising the bodies and public companies of the Shabiat, exploitation and promotion of the local wealth in the Shabiat. Deciding on the matters which are not within the authority of basic popular conferences Peoples Committees, execution of urban planning, environment protection

code, the Shabiat Peoples Conference enjoys extraordinary entity and independent financial status

General People's Congress

It is the general gathering body for all people's conferences, people's committees, unions, associations and professional syndicates and confederation. It is concerned with drafting the laws and decisions issued by the basic people's conferences, choosing the General People's Congress secretariat, questioning it and accepting the resignation of its secretary or any of its members and relieving them from their duties. It also deals with determining the sectors which are managed by general people's committees and determining their duties, choosing the secretary of the general planning board and the secretariat of

A wise libyan man

the G.P.C. secretary, questioning them, accepting their resignation and relieving them from their duties. In addition, it is concerned with choosing the chairman of the Supreme Court and people's court chairman, the prosecutor general, the attorney general, the chairman and members of the people's court at general people's congress. Finally, it deals with accepting their resignations and relieving them of their duties, choosing the people's control organization secretary, the central bank of Libya chairman, questioning them and accepting their resignations and relieving them from their duties.

The Secretariat of the General People's Congress (G.P.C.)

The G.P.C. usually chooses its secretariat among those who were chosen by the people's direct authority. This secretariat is concerned with following up the execution of the laws issued by basic people's conferences, following up the performance of people's committees calling for convening of the G.P.C., calling for convening of unions, associations, and professional confederations, holding joint meetings with other public bodies, organizations or committees, and issuing laws regarding occurrences of basic people's conferences. It also deals with the reviewing and interpretation of laws, regulations and decisions. It is concerned with reviewing treaties and agreements, granting permission to investigate those who were chosen by the G.P.C. and secretaries and members of shabiats people's conferences. It calls for the people's choice and follows its processes. It deals with transferring the cases which require investigation by the control people's committee organization or the general persecutors office, promotion of relations with foreign bodies and organizations on the public and official levels. It is concerned with granting and confiscating the Libyan nationality, and approval of granting political asylum as well as permission to carry merits awarded by foreign bodies to the Jamahiriya citizens.

The Secretariat of the People's Conferences of the Shabiats (Provinces)

This consists of the G.P.C. secretariat and the secretariats of the people's conferences for shabiats. It is concerned with following up of implementation of basic people's conferences decisions and following up of shabiats general people's committee and the affiliated organizations and bodies performance. It also deals with deciding the dates for people's conferences meetings and collecting the drafts of these conferences regarding the agendas. It deals with holding joint meetings with the general peoples committee of shabiat, proposal of laws and regulations, preparation of the financial regulatory projects for the work of secretariats of basic people's conferences and shabiat people's conferences secretariat.

The General Secretariat for Basic People's Conferences

This consists of the G.P.C. secretariat, secretaries of the shabiats people's committees and secretaries of the basic people's conferences. It is concerned with following up of basic people's conferences performance and implementing their decisions, following up the general people's committee and its relevant body performance. It also deals with setting up the regulations for establishing and merging of basic peoples conferences, determining the sectors for which members must be chosen at the level of the basic peoples conference and the conference of the shabiat. Finally, it chooses the Supreme Court councillors and the people's court as well as members of people's prosecution office.

The General People's Committees for Sectors

These are managed and administered by general peoples committee, including the general peoples committees secretaries for sector in the peoples conferences. The G.P.C. chooses the secretaries of the general people's committees for sectors from among those who were directly chosen by the people, or from among any qualified others. The G.P.C. also chooses deputy secretaries for the general people's committees sectors that need deputy secretaries. The general people's committee for the sector has a secretariat which consists of its secretary and relevant secretaries of the sectors. As for the secretariats of general people's committees for sectors which have deputies, their secretariat consists of its general secretary and his deputies. The general people's committees for sectors are concerned with implementing and managing the projects of a special or strategic nature or that which serves more than one shabiat. They also deal with proposing plans, programs and executive procedures of basic peoples conferences decisions as well as proposing budgets and transfers to sectors. Finally it deals with following up organizations and companies interests which fall under its jurisdiction and preparing drafts of legislations concerning the sector.

The General People's Committee

This consists of secretaries of basic conferences people's committee secretaries, secretaries of the shabiats people's committees and secretaries of the general people's committees for sectors.

Walk and discover every corner of Libya

The G.P.C. usually chooses the secretary of the general peoples committee and its deputies those who were directly chosen by the people or among others. The general people's committee is concerned with the execution of laws and decisions issued by basic people's conferences drafted in the G.P.C. proposal of budgets projects, preparing transformation project plans and the general projects and transferring to the planning board proposal of laws drafts in order to submit it to basic people's conferences for discussion. It is also in charge of implementing and managing general and strategic projects, following up the general people's committees for sectors performance and all relevant people's committees. In addition, it deals with supervising and following up manmade river project implementation and its investments as well as promoting foreign investments in Libya and Libyan investments abroad.

The white and soft sand of Libya embraced by a crystal clear Mediterranean

The General People's Committee for Shabiats

This consists of the general people's committee secretary and its deputies, the general people's committees for sectors secretaries, the secretaries of shabiats people's committees. It is concerned with the execution of laws and decisions of basic peoples conferences drafted in the G.P.C. planning the executive programs for partnership system and national service. It also deals with supervising the shabiats people's committees, reviewing the joint matters among shabiats and their coordination and the general people's committees for sectors.

The Secretariat of the General People's Committee

This consists of its secretary and his deputies and the secretaries of the general people's committees for sectors. It is responsible for following up the performance of the aforementioned secretaries of shabiat people's committees

and secretaries of organizations and companies of its concern. It is concerned with the implementation of laws and decisions of basic people's conferences drafted by the G.P.C. and following up the execution of the people's committee performance and the general people's committee for shabiats. It also deals with proposing the law drafts to be submitted to basic people's conferences for discussion, concluding treaties and agreements and international loans and taking the relevant procedures for submitting it to the basic people's conferences for approval. In addition, it is in charge of approving the joint agreements signed between general people's committees and foreign countries, naming secretaries and members of the people's committees for general organizations and companies and naming members of their board of directors. It also grants permission for the abovementioned organizations and companies to conclude contracts and agreements with foreign companies and organizations for project implementation. Finally, it deals with investigating and questioning secretaries and members of peoples committees and punishing them accordingly.

For all those who are interested and follow the people's authority system in the Jamahiriya, you need to know that:

– The collective group is the main decision maker in the

people's authority.

– Decision of the basic people's conference is taken by a majority of opinions and if there are differences in opinions, the different opinions are gathered and the conference decisions then drafted on a majority basis.

– In direct people's choice, if the conference could not agree on the same opinion, the person who got the majority of the members recommendations shall be chosen as the secretary, and the one with less will be deputy or a member as situation necessitates. If two or more persons got the same recommendations for the same position, this position then will be taken in shifts.

– It is not permitted for the person to be both a member of people's committees and a member of secretariats of people's conferences, unions, associations, professional confederations or as such.

GEOGRAPHY

Libya is located in the middle of the Africa's northern coast of the Mediterranean Sea and extends inland to the middle north of Africa. It lies between latitudes 18°-33°N and longitudes 9°E and 25°E. It is bordered by Egypt to the east, Sudan to the southeast, Chad and Niger to the south and Algeria and Tunisia to the west and northwest respectively.

Libya is the fourth largest state in Africa with an area of approximately 1,774,440 sq. km, three times the surface area of France. It largely consists of barren rocky and sandy desert; over 90% of the country is covered by desert. Along the southern border near Chad rises the rugged mountain range of Tibesti, which contains Libya's highest point (3,376m). The Libyan seaboard stretches for 1,774 km along the Mediterranean coast, from Ras Jdayr in the west to Amsaad in the east.

The northwest region, known as Tripolitania, rises from the narrow coastal plain in a series of steps until it reaches the Jafara Plain and the Jabal Nafusah Plateau. Land here rises to between 2,000 and 3,000 feet above sea level. In the northeastern region, known as Cyrenaica, the land rises from the coastal plain to the Green Mountains (Jabal al Akhdar) to a height of just under 3,000 feet.

CLIMATE

Most of the Libyan areas are dry, with a big variation in the temperature. The great desert in the south and the Mediterranean in the north are the main factors for the determination of the country's climate. In the coastal area, the winter season is moderate despite snow falling in some high places sometimes and the temperature in this season is not less than 5 degrees Celsius .(i.e. the west Green Mountain). The summer season is relatively hot and the temperature arrives at its maximum in August which is not over 30 Celsius. There is no rain in summer. The elevation and closeness to sea of some areas (i.e. the West Mountain and Green Mountain) influence the temperature of such areas.

Although the rain is not common in the desert areas, there are sometimes rainstorms and snow there.

Generally, the Libyan climate is moderate and varied, which makes it attractive for tourism.

Regarding the climatic seasons and tourist seasons the country may be divided into four distinctive areas:

Sunset in Libya, just a delightful experience

Tourism seasons in the light of the geographical areas

Geographical Area	Main tourist season
Coastal area	All year
Eastern area	From June to September
Western area	From May to September
Desert area	From October to the middle of March

LANGUAGE

Arabic is spoken all over the country, as it is the official language of Libya. Some of the elders in the north of the country speak some Italian, but English is the second language.

Religion

Islam is the state religion, and all Libyans are Muslims. As Libyans are Sunni Muslims almost across the board, in general they are conservative without being fundamentalist in their outlook.

Education

The country has now one of the highest literacy rates in Africa. Some 97% of the adult population is literate. Public education in Libya is free and compulsory for children ages 6 to 15. Arabic is the language of instruction. Great strides have been made in education. Literacy rates, as well as the number of classrooms, teachers, university students and female students have risen significantly. Libyans have a variety of options for higher education, including universities and various technical, vocational and agricultural institutes. The University of Garyounis in Benghazi and Al-Fateh University in Tripoli are the best known. Libyan law provides that education is a right and duty of all Libyans. It is compulsory from the age of six until the end of secondary school. The state guarantees this right through the establishment of schools, institutes and universities, and of pedagogical and cultural institutions in which education is free. The establishment of pri-

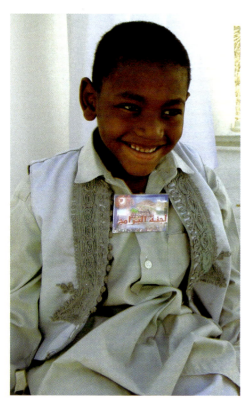

Enjoy the smile of a libyan kid

has increased to 87.4% in accordance to the 2002-2003[1] economic and social census results. It is expected to increase again after seeing the results of general population census that will be carried out in 2005, especially if we knew that the illiteracy ratio is concentrated in the elderly people category. The ratio of the students to the total population in accordance to the data of 2002-2003 censuses was 23%[2].

Health

There are no special health precautions before coming to the country as there are no known sicknesses that would be dangerous to travellers coming to Libya. However, it is recommended that travellers bring U.V. protection cream and hats to protect themselves from the sun when travelling to the coast or to the desert. There are many hospitals owned by the government and many private clinics in Tripoli and other big cities. Living standards in Libya are generally good.

Villas and apartments are spacious and usually air-conditioned for expatriates. Most European foodstuffs and other consumer products are available in the supermarkets and stores at reasonable prices. Local farm produce such as dairy products, fruit and vegetables is found in abundance and is fairly inexpensive.

Driving licence & permits

Concerning the documentation, foreign driving licences are valid for three months. After that a Libyan licence must be obtained. When staying longer than three months every foreigner driving in Libya must obtain the local driving licence. The simplest way of getting it is to come to the country with an international driving licence and, with the help of a local contact, try to get a Libyan driving licence if possible. Car rental agencies should help in getting all the necessary documents. A special aspect of Libya concerning foreign drivers is that it was one of

vate educational institutions is also regulated by law. The state is particularly anxious to enhance the physical, intellectual and moral development of the youth. There are also good international schools that cater to the needs of children of the expatriate community

Since 1969 when the revolution took place, Libya has witnessed an accelerated development in the educational process that was in line with the other economical and social developments that the country witnessed since then. The data of the last population census made in 1995 point out a decline in the illiteracy percentage among Libyans, and an increase in the ratio of people who can read and write from 67.62% in 1984 to 81.31% in 1995. This ratio

SOME INDICATIVE PRICES

Pack of cigarettes: $1.75

One night at a 4-5 star hotel: between $70 and $250

Gas: close to 5$ for a full tank

Rent a house: between $1000 and $2000 monthly

Taxi fare: $20 per hour

DVD/VCD rental: $0,70 per day

Driver's salary: between $150 and $200 per week

Housekeeping: $5 per hour

A stamp from the libyan post

the first and few countries with Russia that actually identifies any car owned by a foreigner through a special number located in a blue frame showing its correspondent country number.

Visa Requirements & Visa Extensions

To enter into the Jamahiriya it is necessary for all foreigners to obtain entry visas which are granted by the Libyan People's Bureau abroad or by such an office that represents the Jamahiriya abroad. Also, tourists coming into Libya in groups can be granted entry visas at official points of arrival, provided that they have an invitation

from a recognized Libyan travel agency or tour operator; and this also applies to investors who have been invited by an official investment authority.

Non-Arab and non-residing foreigners can obtain entry visas in cases where they have an official invitation to be addressed by an organization inside Libya that is desirous or involved in the procedure of entry visa issue such as tourist agencies and companies. In addition they can enter the Jamahiriya only with a valid travel document from their countries.

- Entry fees decided by the Tourist Licensing and Inspection Administration at the General People's Committee for

It is essential to taste delicious libyan tea enjoying the atmosphere of a traditional desert tent

Tourism are LD 5,000.

Points of Arrival:

- Seaports: Tripoli Seaport, Benghazi Seaport.
- International Airports: Tripoli International Airport and Benina Airport.
- Land entry points:
• Ras-Ejdair entry point.
• Emsaed entry point.

MAIN INTERNATIONAL AIRPORTS IN LIBYA

Tripoli International Airport

Tel: + 218 - 21- 3609777
Fax: + 218 - 21 - 3609696

Benghazi International Airport

The number of passengers to and from Tripoli International Airport in 2002 was 1,230,942 and in 2003, the number of passengers was 1,450,220. The number of passengers passing through Benghazi Airport in 2002 was 564,614 and in 2003 it was 660,450.

The main domestic airports are mainly established for tourist purposes and local public transport.

ARRIVING AT / LEAVING THE AIRPORT

Tripoli International Airport is located approximately twenty minutes from the city centre. There are two ways to and from the airport by car. The best is using the new airport road which is a well-paved four-way highway. The second option is to use the "old airport way", a two-way road starting in Benghazi. The best way is to use the local taxis which will take you from the airport to any of the main hotels for a rate of approximately US$10. It is always rec-

Libya has a wide variety of traditional handicrafts available in most cities

ommended to previously arrange the price with the driver.

AIRLINES

Air Algerian
E.megharief St.
Tel: + 218 - 21 - 444017

Afriqiyah Airways
Waha Building-Omar Almokhtar St.
P.O. Box 83428 Tripoli
Tel: + 218 - 21 - 4449734

Air Malta
Bourj Alfatah, Office 24 ground floor
Tel: + 218 - 21 - 3350579 / 3350578

Alitalia
Dat Al Emad Bldg Tower 3 first floor
Tel: + 218 - 21 - 3350296 / 3350279

Austrian Airlines
Dat Al Emad Bldg Tower, 3 Level 6
Tel: + 218 - 21 - 3350241 / 3350243

British Airways
Al Fateh Tower 1 Level 19
Tel: + 218 - 21 - 3351277

Egypt Airlines
Awel September St.
Tel: + 218 - 21 - 3335781 – 3335782

Emirates Airlines
Al Fateh Tower 1 Level 10
Tel: + 218 - 21 - 3350592 / 3350595

Lufthansa
Dat El Emad Bldg

Tel: + 218 - 21 - 3350375

Pakistan Airlines
Dat El Emad Bldg, Tower 5 ground floor
Tel: + 218 - 21 - 3350448 / 449

Royal Jordanian
E.megharief St.
Tel: + 218 - 21 - 4446695 - 3334685

Royal Maroc Airlines
Dat El Emad Bldg, Tower 5 Level 6
Tel: + 218 - 21 - 3350111 / 3350112

Swissair
Bourj Alfatah, Tower 1 Level 18
Tel: + 218 - 21 - 3350052 / 3350053

Syrian Arab Airlines
Baladia St.
Tel: + 218 - 21 - 4446715 / 4446716

Tunis Air
Awel September St.
Tel: + 218 - 21 - 4446499 / 3336303

Turkish Airlines
Bourj Alfatah, Tower 2 Level 16
Tel: + 218 - 21 - 3351351 / 53

K.L.M.
Bourj Alfatah, Tower 1 Level 16
Tel: + 218 - 21 - 3350018 / 3350019

Qatar Airlines
Bourj Alfatah, Tower 1 Level 10
Tel: + 218 - 21 - 3351810 / 816

Sudan Airways
Bourj Alfatah, Tower 1 Level 12

Enjoying the typical green tea in front of the sea

Tel: + 218 - 21 - 3351330 / 31

Malev Hungarian Airlines
Bourj Alfatah, Tower 1 Level 19
Tel: + 218 - 21 - 3351245

Jat Yugslavian Airlines
Bourj Alfatah, Tower 1 Level 17
Tel: + 218 - 21 - 3351300

Pia Paktanian International Airlines IA
Dat El Emad Bldg
Tel: +218 - 21 - 3350448

Hemus Air
Baladiyah St. 114
Tel: + 218 - 21 - 4445560 / 3339748

Libyan Arab Airlines
Haiti Street
Tel: + 218 - 21 - 3614824 / 3614102

BUSINESS WORKING HOURS

Government offices are generally open between 8am and 2.30pm from Saturday to Thursday. Business hours in private offices tend to be from 8am to 1pm and from 4pm to 8pm from Saturday to Wednesday and from 8am to 2pm on Thursday. Shopping hours are from 9am to 2pm and 4.30pm to 8.30pm.

BANKS

Agricultural Bank
Tripoli
Tel: + 218 - 21 - 3333541 / 44
Fax: + 218 - 21 - 3330927

Aman Commerce and Investment Bank
Dat Al Emad Bldg Tower 3, 1st floor
P.O. Box 91271, Tripoli
Tel: + 218 - 21 - 3350219 / 16
Fax: + 218 - 21 - 3350387

BACB
Al-Fateh Towers, Tripoli
Tel: + 218 - 21 - 3351730 / 3351731
Fax: + 218 - 21 - 3351732

Bank of Commerce & Development
P.O. Box 5045, Benghazi
Tel: + 218 - 61 - 9097300 / 218-61 97220

Bank of Valetta
Al Fateh Towers, Tripoli
Tel: + 218 - 21 - 3351661 / 2

Central Bank of Libya
P.O. Box 1103, Tripoli
Tel: + 218 - 21 - 333591 - 99

Development Bank
P.O. Box 3180, Tripoli
Tel: + 218 - 21 - 4802044 / 5

Libyan Arab Foreign Bank
Dat El Imad Bldg., Tripoli
Tel: + 218 - 21 - 3350155 - 60
Fax: + 218 - 21 - 3350024

National Commercial Bank
Tripoli tel: + 218 - 21 - 4441168 / 9
Benghazi tel: + 218 - 61 - 91141 / 49132
Sebha tel: + 218 - 71 - 39128 / 20185
Al Baida tel: + 218 - 84 - 22411 / 23386

Sahara Bank
P.O. Box 270, Tripoli
Tel: + 218 - 21 - 3330724 / 3339265
Fax: + 218 - 21 - 3337922

Umma Bank
P.O. Box 685, Tripoli
Tel: + 218 - 21 - 3331195 / 3332888
Fax: + 218 - 21 - 3330880

Wahda Bank
P.O. Box 3427, Tripoli
Tel: + 218 - 21 - 3336513
Fax: + 218 - 21 - 3337090

FOREIGN EXCHANGE

Visitors can exchange and buy foreign currencies at some of the banks mentioned above, there is also the possibility of exchanging currency on the black market but it is not recommended as the difference in the exchange rate is not worth it and is not as safe as the banks. Foreign currency can be imported freely provided it is declared on arrival. A minimum of US$500 or equivalent in a readily acceptable currency must be exchanged upon arrival. Export of foreign currency is limited to the amount declared on import; exchange back is subject to having spent at least US$50 per day during the stay. The import and export of local currency is prohibited.

MONEY & CREDIT CARDS

In Libya the currency is the Libyan dinar (LD). One Libyan dinar equals to 1000 dirhams. Currency notes are in denominations of LD 20, 10, 5 and 1, and 500 and 250 dirhams. Coins are in denominations of 100, 50, 25, 20, 10, and 5 dirhams. Exchange rate indicators are against sterling and the US Dollar. The following figures are included as a guide to the movements of the Libyan dinar against sterling and the US dollar:

Date	Sep'99	May'99	Oct'99	Nov'99	Mar'04	April'04	May'04	June'04
£1.00=	0.64	0.72	0.74	0.78	2.36	2.39	2.39	2.41
$1.00=	0.38	0.45	0.45	0.47	1.30	1.31	1.33	1.31

COMMUNICATIONS

The Libyan government owns and operates both the fixed-line and mobile network systems. The postal system is also nationalised with post office box and PTT facilities in all the large towns.

International postal services to Libya can be slow. Average time for airmail is around 7 to 12 days. Surface mail can take considerably longer. There is no guarantee that mail will be delivered, therefore using a good courier service for delivery of business correspondence is recommended. Yamama International Couriers Ltd operations are focused on the key areas of world trade in Europe, Asia, and North and South America. Yamama International Couriers Ltd also has strong domestic networks in Australia. DHL World-wide Express maintains offices around the world in over 200 countries. Courier companies, such as DHL, now provide more reliable delivery times for packages and mail.

Internet access in Libya is becoming increasingly more popular, with the emergence of internet cafes, which are the

CITY	AREA CODE NO.
ZUARA	025
SABRATAH	024
TRIPOLI	021
MISURATA	051
SIRT	054
JDABIYA	064
BREGA	0656
TAWERGHA	0522
JADU	044
DERNA	081
ZAWIA	023
GHADAMES	0484
KABAO	0481
WADDAN	0581
SEBHA	071
TOBRUK	087
TOLMETHA	0685
HUN	057
SHAHAT	0851
OJLAH	0653
BENGHAZI	061
BAYDA	084
TARHUNA	0325
RAS LANOUF	0527
KHOMS	031
GHARIAN	041
AL MARJ	067
YEFREN	0421

favourite pass-time of young Libyans. There is now a local service provider and service companies are now opening e-mail accounts. Most companies now have accounts, and private access from home is also available. It is anticipated that Libya will see a similar growth in cyber technology as seen in other North African markets. However it will be a while before e-commerce is a popular method in which to do business. Even when this is the case, companies should ensure that they have regular face-to-face contact if they wish to ensure continued long-term business in Libya.

Mobile phone technology is limited but developing fast. There are currently roaming service agreements in Libya so foreign mobile phones work. Almost everyone in business has a mobile phone and they tend to be used far more for business than they are in Europe.

The international code for Libya is 218, and main internal area codes are as follows:

- Tripoli: 21
- Benghazi: 61
- Tobruk: 87
- Tripoli International Airport: 22

Fax facilities are available at the larger hotels, and from 24-hour PTT offices.

Time

Libyan Standard Time is two hours ahead of GMT and 7 hours ahead of US Eastern Standard Time. There is no time difference in summer with Spain or France but there is one hour difference in winter (+1). As in many other Muslim countries, public and private sectors are closed on Fridays and some might be closed on Thursday afternoon too.

Calendar

Dates are usually written in the order of day, month and

A view of Tripoli combining both leisure and business, a perfect eBizguides situation

year, however, other methods are also used for the Gregorian calendar dates. The Islamic calendar is also used, in particular for religious holidays and other occasions.

Libyan Months of the year

1- January	Ai-Nnar
2- February	An – Nuar
3 – March	Al – Rabii
4 – April	Al- Tair
5 – May	Al – Maa
6 – June	Al – Sayf
7 – July	Naser
8 – August	Hanibal
9 – September	Al – Fateh
10 – October	At-Tumur
11 – November	Al – Hareth
12 – December	Al – Kanoon

Public Hollidays

Libya observes all the main Muslim festivals, with some

additional national holidays. Friday is the official day of rest.

Libyan national holidays are as follows:

National Holidays

Declaration of the Authority of the People	March 2nd
Evacuation of the British Forces	March 28th
Evacuation of the American Forces	June 11th
Egyptian Revolution	July 23rd
First of September (Al-Fateh)	
Revolution	September 1st
Evacuation of the last Italian Settlers	October 7th

Religious Holidays

Islamic (Lunar) New Year	1st Moharran
The Prophet's Sacred Birthday	12th Rabii Al-Thani
Eid Al-Fitr (end of Ramadan)	1st to 3rd Shawal
Eid Al-Adha	9th to 12th Dhul-Hijja

Getting around

By air:

A number of major international airlines fly to Libya regularly, including British Airways, Lufthansa, Swissair and Alitalia. Libyan Arab Airlines (LN) with AL-Buraq Airlines provide fast and frequent internal services between Tripoli, Benghazi, Sebha, Al Bayda, Mersa Brega, Tobruk, Misurata, Ghadames and Al Khufrah. They also offer an hourly shuttle between Tripoli and Benghazi.

By road:

The main through road follows the coast from west to east. Main roads are Al Qaddahia–Sebha, Sebha–Ghat, Tripoli–Sebha, Agedabia–Al Khufrah, Garian–Jefren, Tarhouna–Homs, Mersa Susa–Ras, Hilal–Derna and Tobruk–Jaghboub. Since 1969, signposts other than those in Arabic script have been prohibited; signs and house numbers are, in any case, rare outside the main towns. Petrol is available throughout Libya, and is currently about half the price of that in Britain. There are no reliable town maps. The

Bus is one of the most common ways of transportation inside Tripoli

quality of servicing is generally poor by European standards, as is the standard of driving. Traffic drives on the right.

Car rental: Self-drive cars are available in Tripoli and Benghazi. National driving licences are valid for three months. After that a Libyan licence must be obtained.

Buses are available around town, they can be recognisable as they are painted white and yellow, you can go inside those buses along the route but it can be quite difficult for the foreign traveller to use them. There is a bus service between Tripoli and Benghazi and other cities.
A minibus service operates from Tripoli to other major towns.

Taxis in Libya are in very common use, it is easy for a traveller to look for taxis at the main door of the hotel. Those taxi drivers waiting down at the hotel are the ones who speak English in town. In Tripoli there are two kinds of taxis: the black and white ones and the ones in yellow and white. It is quite easy to recognise them as the plate of the car is always yellow, but there is a difference between them:
The black and white taxis are providing the kind of service that Europeans and Americans for example, are used to. They will take their clients anywhere around the city for a certain fare. Some of them might have a taxi fare digital indicator, others will not so you need to arrange the amount to be paid at the destination with the driver before the journey. It is also possible to arrange for those drivers to take you outside Tripoli, but then the fare will get bigger according to the distance, it is better to arrange the price before leaving the city.

The other taxis in yellow-white are different because they are seven seater cars that work like buses along the same street, so they won't take a single passenger but they will take as many people as they can at any point of the itinerary they are covering. Those taxis are cheaper but won't be really helpful to you in cases that you need to go to a certain address that may be out of the itinerary; also the driver may not speak English.

In any case, taxis are quite safe and quite professional in Tripoli. However, for a period of more than two to three days stay it is recommended to reach an agreement with a driver staying near one of the main hotels in town. Please see the business resources section for contact information on English speaking drivers.

LIBYAN EMBASSIES ABROAD

Azerbaijan

Embassy of The Great Libyan Jamahiriya
Husein Agawed Quarter, No.520, Beldnn.No.20, P.O. Box 9
Tel: + (944) 938548 - 932365
Fax: + (944) 989770 - 981247

Algeria

Embassy of The Great Libyan Jamahiriya
Shihk Bashir Ibrahimy, Alabyar, Algiers P.O. Box 97
Tel: + (213) 921502 / 922589
Fax: + (213) 924687

Austria

Embassy of The Great Libyan Jamahiriya
Blaasstrasse 33, 1190 Vienna, Austria
Tel: + (43) 3677639
Fax: + (43) 3677601

Bahrain

Embassy of The Great Libyan Jamahiriya
Villa 787 Street 3315, Compound 333, Um Alhusm, Manama, Bahrain
Tel: + (973) 722252
Fax: + (973) 722611

Belgium

Embassy of The Great Libyan Jamahiriya

Avenue-Victoria 28, Brussels, Belgium

Tel: + (32) 6492113 / 6493737

Fax: + (32) 6409076 / 645389

Bosnia Herzegovina

Embassy of The Great Libyan Jamahiriya

Tahliti Sukak 17 Sarajevo, Bosnia Herzegovina

Tel: + (387) 660387

Fax: + (387) 4444220

Brazil

Embassy of The Great Libyan Jamahiriya

Shis Q1-15 Chacara – 26 Laga Sul,

70462900 Brasilia, Brazil

Tel: + (55) 248716 / 248728

Fax: + (55) 2480598

Bulgaria

Embassy of The Great Libyan Jamahiriya

Mlado St. 1 Bould Andrei Sakharov,

Residence No.1 Sofia 1784, Bulgaria

Tel: + (359) 9743156 / 9743556

Fax: + (359) 9743273

People's Republic of China

Embassy of The Great Libyan Jamahiriya

3 Dong Liu Jie Sanli Tun, Peking, China

Tel: + (86) 6532980 / 65323666

Fax: + (86) 65323391

Cuba

Embassy of The Great Libyan Jamahiriya

Calle 7 Ma No. 1402 Esquina, A-14

Miramar Playas, La habana

Tel: + (53) 2040150 / 204892

Libya is a young country: an important proportion of its citizens are teenagers

Fax: + (53) 2042991

Cyprus

Embassy of The Great Libyan Jamahiriya
14 Estlas Street No. 23669, Nicosia
Tel: + (357) 22460055
Fax: + (357) 22452710

Czech Republic

Embassy of The Great Libyan Jamahiriya
NA Baste SV JIRL 57 16000, Prague
Tel: + (420) 341929 / 325325
Fax: + (420) 33322173

Denmark

Embassy of The Great Libyan Jamahiriya
Rosen Vaengets Hodvej 4DK-2100, Copenhagen, Denmark
Tel: + (45) 35263611 / 35263664
Fax: + (45) 35265606

Egypt

Embassy of The Great Libyan Jamahiriya
Al-Saleh Ayoub Street, Al-Zmalek, Cairo
Tel: + (20) 1 7357862 / 7358216
Fax: + (20) 1 7359319 / 7356072

France

Embassy of The Great Libyan Jamahiriya
2 Rue Charles Lamoureux 75116, Paris, France
Tel: + (33) 45534070 / 47047160
Fax: + (33) 47559625

Germany

Embassy of The Great Libyan Jamahiriya
Libysche Araisdghe Volks-Jamahyria, Grosse
Sozialists CGE Volksuro, BeethovenAllee
12A 53173 Bonn
Tel: + (49) 8200911 / 8200964
Fax: + (49) 364260

Hungary

Embassy of The Great Libyan Jamahiriya
1143 Budapest 14 Stefania UT 111 Hungary, P.O. Box
1411P.F. 73 Budapest
Tel: + (36) 3649336 / 3649332
Fax: + (36) 3649330 / 3449334

India

Embassy of The Great Libyan Jamahiriya
22 Golf Links New Delhi 110003, India
Tel: + (91) 1 4697717 / 4697771
Fax: + (91) 1 4633005

Indonesia

Embassy of The Great Libyan Jamahiriya
JL-Pekalongan 24 Menteng, Jakarta 10310, Indonesia
Tel: + (62) 21 335308 / 335754
Fax: + (62) 21 35726

Iran

Embassy of The Great Libyan Jamahiriya
Marem Street Building No.2, Tehran, Iran
Tel: + (98) 21 2242797 / 2201677
Fax:+ (98) 21 2236649

Italy

Embassy of The Great Libyan Jamahiriya
Via-Baracchini No.7, Milano, Italy
Tel: + (39) 2 86464285
Fax: + (39) 2 874504

Italy

Embassy of The Great Libyan Jamahiriya
Via Nomentana No.365, Roma, 00162 Italy
Tel: + (39) 6 86320951 / 9
Fax: + (39) 6 86205473

Italy

Embassy of The Great Libyan Jamahiriya
Viale Dlla Liberta 171, Palermo, 90134 Italy

Tel: + (39) 91 343930 / 343931
Fax: + (39) 91 343932

Japan

Embassy of The Great Libyan Jamahiriya
Tokyo 150-0034, Japan
Tel: + (81) 3 34770701 / 3
Fax: + (81) 3 34640420

Kazakhstan

Embassy of The Great Libyan Jamahiriya
Almaaa Kazakhistan, 10 Melnichnaya St.,
480100 Kazakhistan.
Tel: + (7) 453519 / 542748 / 542745
Fax: + (7) 608536

Kingdom of Jordan

Embassy of The Great Libyan Jamahiriya
Al-shemsan Wadi Hanfra, Jubran Khalil Jubran St. No.7,
Amman, Jordan
Tel: + (962) 6 5693101 / 5693102
Fax: + (962) 6 5693404 / 5662956

Kingdom of Saudi Arabia

Embassy of The Great Libyan Jamahiriya
Jeddah Saudi Arabia, Alhamara Hay-Alandalus, Baridi St,
beside Siqua Alhamraa, PO Box 574, Saudi Arabia
Tel: + (966) 2 4625302 / 4610684
Fax: + (966) 2 4610685

Kuwait

Embassy of The Great Libyan Jamahiriya
Aldayaa Sector 1 Street No.14, Villa No.3, Kuwait
Tel: + (965) 2575183 / 2575184
Fax: + (965) 2575182

Lebanon

Embassy of The Great Libyan Jamahiriya
Beirut, Lebanon, Abdullah Almashgouf St., Beirut
Saqia Al-janzeer (Furdan) Building No.(149), PO Box 3153.

Tel: + (961) 806314 / 785859
Fax: + (961) 20869 / 22181

Madagascar

Embassy of The Great Libyan Jamahiriya
Antanarivo PO Box 116, Madagascar
Tel: + (261) 2 21885 / 21892
Fax: + (261) 2 z25672

Malaysia

Embassy of The Great Libyan Jamahiriya
No. 6 Jalan Madge off Jalan, Uthant 55000 Kuala Lampur,
Malaysia
Tel: + (60) 3 21411293 / 21482112
Fax: + (60) 3 21413549

Malta

Embassy of The Great Libyan Jamahiriya
Notabile Roao Balzan, Malta
Tel: + (356) 483837 / 447883
Fax: + (356) 483939

Morocco

Embassy of The Great Libyan Jamahiriya
2nd Floor B., Markeb Aljass Aljayesh Almalaky,
Casa Blanca, Casa Blanca, Morocco
Tel: + (212) 631871 / 72
Fax: + (212) 631874 / 631878

Mozambique

Embassy of The Great Libyan Jamahiriya
Rua Dereira 274, P.O. Box 4434, Maputo, Mozambique
Tel: + (258) 1 492662
Fax: + (258) 1 490140 / 492450

Nigeria

Embassy of The Great Libyan Jamahiriya
Lagos, Plot C 11 Ligali Ayonrinest Victoria, Victoria Island
P.O Box 435, Abuja, Nigeria
Tel: + (234) 1 610107 / 610110

Fax: + (234) 1 5231570 / 5231571

Panama

Embassy of The Great Libyan Jamahiriya
Avenida Bilbao con calle, 32 Esquina Frente Edificio, Atalya,
Panama
Tel: + (507) 2273342 / 2273365
Fax: + (507) 2273886

Philippines

Embassy of The Great Libyan Jamahiriya
1644-Dasmarines-Corner Mabolo, Dasmarines Village
Makati City, Philippines
Tel: + (63) 2 8177331 / 8177332
Fax: + (63) 2 8177337

Poland

Embassy of The Great Libyan Jamahiriya
Ul Kryniczna 2-4-03-934, Warsaw, Poland
Tel: + (48) 22 6174822 / 61744883
Fax: + (48) 22 6173029 / 6175091

Portugal

Embassy of The Great Libyan Jamahiriya
Av. Das Descobertas 24, Restelo 1400, Lisbon, Portugal
Tel: + (351) 1 3016301 / 02
Fax: + (351) 1 3012378

Qatar

Embassy of The Great Libyan Jamahiriya
Sulta Aljadeeda Area No.23, Square No. 3, P.O. Box 574 Al
Dawha, Qatar
Tel: + (974) 4932556 / 4932557
Fax: + (974) 4839407

Romania

Embassy of The Great Libyan Jamahiriya
Bd Ana Ipaescu No.15, Bucharest, Romania
Tel: + (40) 1 2125705 / 2127832
Fax: + (40) 1 3120232

Russia

Embassy of The Great Libyan Jamahiriya
Mosfil Movaskaya Str 38, 00940 Moscow, Russia
Tel: + (7) 095 1430354 / 1437700
Fax: + (7) 095 9382162

Slovakia

Embassy of The Great Libyan Jamahiriya
Revova Ulica 45/81102, Bratislava, Slovakia
Tel: + (421) 7 54410324
Fax: 54410730

South Korea

Embassy of The Great Libyan Jamahiriya
4/5 liannman Dong Ongasan Ku, P.O. Box 8418, Seoul,
Korea
Tel: + (82) 2 7976001 / 5
Fax: + (82) 2 7976007

Spain

Embassy of The Great Libyan Jamahiriya
C/ Pisuerga No. 12 28002, Madrid, Spain
Tel: + (34) 915644675 / 914114743
Fax: + (34) 915643986

Sudan

Embassy of The Great Libyan Jamahiriya
Mashtal Street Alryad, Al Khartoum, Sudan
Tel: + (249) 11 222085 / 222457
Fax: + (249) 11 727319

Sweden

Embassy of The Great Libyan Jamahiriya
Vahalla Vagen 74
P.O. Box 10133, Stockholm, 10055 Sweden
Tel: + (46) 8 143435 / 38
Fax: + (46) 8 104380

Switzerland

Embassy of The Great Libyan Jamahiriya

GENERAL INFORMATION ABOUT LIBYA

Tavel Weg 2 Ch-3006 Bern, Switzerland

Tel: + (41) 22 3513076 / 77

Fax: + (41) 22 3511325

Tunisia

Embassy of The Great Libyan Jamahiriya

48 Mukarar Nahj Ghara Jwan-Mtwak Feel, Tunis, Tunisia

Tel: + (216) 1 781913 / 780866

Fax: + (216) 1 795338

Tunisia

Tunisia Libyan Consulate 2 Yousef El Roisi Streif

Tel: 7187512 / 71875163

Fax: 71875162

Turkey

Embassy of The Great Libyan Jamahiriya

Janah Jadsy Raqan 60 shanqaya, Ankara, Turkey

Tel: + (90) 312 4381114 / 4381110

Fax: + (90) 3124403862

Turkey

Embassy of The Great Libyan Jamahiriya

Libya Haik Burosu Cinnah Caddesi, No 60 Cankaya, Ankara,Turkey

Tel: + (90) 312 4381110 / 4381114

Fax: + (90) 312 4419523

An oasis waiting to be discovered by you

Ukraine

Embassy of The Great Libyan Jamahiriya
6 Ovruchskaya St. 04050, Kiev, Ukraine
Tel: + (380) 44 2386071 / 2386070
Fax: + (380) 44 2386068

United Arab Emirates

Embassy of The Great Libyan Jamahiriya
Dubai UAE
Tel: 450030
Fax: 450033

United Kingdom

Embassy of The Great Libyan Jamahiriya
61-62 Ennismore Gardens, London SW 7 1NH, England
Tel: + (44) 020 - 75896120
Fax: + (44) 020 - 75896094

United States of America

Embassy of The Great Libyan Jamahiriya
The Permanent Mission Of The Peoples Libyan Arab Jamahiriya To The United Nations 309 East 48th Street, New York N.Y. 10017, USA
Tel: + (1) 212 7525775
Fax: + (1) 212 5934787

Venezuela

Embassy of The Great Libyan Jamahiriya
3Ra Avenide A1 10 ma transveral Los Leones, Altaira Tamira Ente 9, Caracas, Venezuela
Tel: + (58) 212 2639698 / 2613953
Fax: + (58) 212 2617271

Vietnam

Embassy of The Great Libyan Jamahiriya
A3 Van Phuch, Hanoi, Vietnam
Tel: + (84) 4 8463503 / 8453379
Fax: + (84) 4 8454977

Yugoslavia

Embassy of The Great Libyan Jamahiriya
St. Merko To Mecha 6, Belgrade, Yugoslavia
Tel: + (381) 663445 / 668253
Fax: + (381) 3670805

Zimbabwe

Embassy of The Great Libyan Jamahiriya
124 Harare street, P.O. Box 4310, Harare, Zimbabwe
Tel: + (263) 4 774885 / 774886
Fax: + (263) 4 774888

FOREIGN EMBASSIES IN LIBYA

Embassy of Afghanistan

Mothafar Alaftas Street P.O. Box 4245, Tripoli
Tel: + 218 - 21 - 4841441
Fax: + 218 - 21 - 4841443

Embassy of Algeria

Tripoli
Tel: + 218 - 21 - 4447072 / 4447043
Fax: + 218 - 21 - 3334631

Embassy of Argentina

Mothafar Alaftas Street P.O. Box 932, Tripoli
Tel: + 218 - 21 - 4834956
Fax: + 218 - 21 - 4840928

Embassy of Austria

Khalid Ibn Al Walid Street – Dahra, Tripoli
Tel: + 218 - 21 - 4443393
Fax: + 218 - 21 - 4440838

Belgian Embassy

5th Floor Dhat Al Imad Tower 4, Tripoli
Also accredited to: USA interests
Tel / fax: + 218 - 21 - 3333771 / 3333660

GENERAL INFORMATION ABOUT LIBYA

Embassy of Benin

Ghout Al Shaal, Tripoli

Tel / fax: + 218 - 21 - 4837663

Embassy of Bosnia and Herzegovina

Ben Ashour Street, Tripoli

Tel: + 218 - 21 - 4776442

Fax: + 218 - 21 – 4774327

Embassy of Brazil

Ben Ashur Street P.O. Box 2270, Tripoli

Tel: + 218 - 21 - 3614896 / 3614894

Fax: + 218 - 21 - 3614895

British Embassy

Al Fateh Tower 24th Floor Sharia Al-Shat, P.O. Box 4206, Tripoli

Tel: + 218 - 21 - 3351084

Fax:+ 218 - 21 - 3351082

Contact person: Chris Rampling

http://www.britain-in-libya.org

Embassy of Bulgaria

Tripoli

Tel: + 218 - 21 - 3609988

Fax: + 218 - 21 - 3609990

Embassy of Burkina-Faso

Tripoli

Tel: + 218 - 21 - 4771221

Fax: + 218 - 21 - 4778037

Gherian is famous for its shops on the side of the road

Canadian Embassy

Al Fateh Tower 7th Floor, Tripoli
Tel: + 218 - 21 - 3351633
Fax: + 218 - 21 - 3351630

Embassy of Chad

Al Dahra, Tripoli
Tel: + 218 - 21 - 4443955
Fax: + 218 - 21 - 4444765

Embassy of China

Gergaresh - Hai Al Andalous P.O. Box 6310, Tripoli
Tel: + 218 - 21 - 4833193
Fax: + 218 - 21 - 4831877

Embassy of Cuba

Tripoli
Tel / fax: + 218 - 21 - 475216

Embassy of Cyprus

Al Thel Street Ben Ashour Road, Tripoli
P.O. Box 3284 Central Post Office
Tel: + 218 - 21 - 3351400

Embassy of the Czech Republic

Ahmed Lutfi Street Ben Ashour Area, P.O. Box, Tripoli
Tel: + 218 - 21 - 3611555
Fax: + 218 - 21 - 3615437

Embassy of Denmark

Ben Ashour Street, Tripoli
Tel / fax: + 218 - 21 - 3607131 / 3607131

Embassy of Finland

P.O. Box 2508, Tripoli
Tel: + 218 - 21 - 605117 / 3338057
Fax: + 218 - 21 - 3341796

Embassy of France - Main Office

Beni El Amar St. - Hay El Andalous Area, P.O Box 312,Tripoli

Tel: + 218 - 21 - 4773807
Fax: + 218 - 21 - 4778266
Opening hours, From Sunday to Tuesday,
From 8 a.m. to 2.30 p.m.
http://www.ambafrance-ly.org/

Embassy of France - Commercial Section

15th Floor Dhat Al Imad Tower 1, Tripoli
Tel / fax: + 218 - 21 - 3350340

Embassy of France - Cultural Section

Karatchi Street, Tripoli
Tel / fax: + 218 - 21 - 3375567

Embassy of Germany

Hassan Al Mashari Street, Tripoli
Tel: + 218 - 21 - 4448552
Fax: +218 21 4448968

Embassy of Greece

18 Jalla Bayar Street, Tripoli
Tel: + 218 - 21 - 3336689
Fax: + 218 - 21 - 3341796

Embassy of Guinea

Tripoli
Tel: + 218 - 21 - 4772793
Fax: + 218 - 21 - 3609990

Embassy of Hungary

Tripoli
Tel: + 218 - 21 - 3618218
Fax: + 218 - 21 - 3618220

Embassy of India

16/18 Mahmoud Shaltout Street P. O. Box 3150, Tripoli
Tel: + 218 - 21 - 4441835 / 4441836
Fax: + 218 - 21 – 3337560

Embassy of Iran
Al Jomhorya Street, Tripoli
Tel: + 218 - 21 - 3615685
Fax: + 218 - 21 - 3611674

Embassy of Iraq
Gurgy Road, Tripoli
Tel: + 218 - 21 - 4770003 / 4770787

Embassy of Italy - Main Office
1 Uaharan Street, Tripoli
Tel: + 218 - 21 - 3334131
Fax: + 218 - 21 - 3331673

Embassy of Italy - Consulate General Benghazi
5 Omar Ibn El Aas Street, Benghazi
Tel / fax: + 218 - 61 - 3334131

Embassy of Japan
That Al-Emad, Tripoli
Tel: + 218 - 21 - 4781043 / 4781041
Fax: + 218 - 21 - 4781044

Royal Jordanian Embassy
Ben Ashour Street, Tripoli
Tel: + 218 - 21 - 3614761
Fax: + 218 - 21 - 3614762

Embassy of the Republic Democratic of Korea
Ben Ashoor Street, Tripoli
Tel: + 218 - 21 - 4831323
Fax: + 218 - 21 - 4831324

Embassy of Kuwait
Ben Ashoor Street, Tripoli
Tel: + 218 - 21 - 4440282
Fax: + 218 - 21 - 607053

Embassy of Lebanon
Ben Ashoor Street, Tripoli
Tel: + 218 - 21 - 3615744
Fax: + 218 - 21 - 3611740

Embassy of Malaysia
Gergaresh - Abu Nawas - P.O. Box 6309, Hai Al Andalus, Tripoli
Tel: + 218 - 21 - 4830854
Fax: + 218 - 21 - 4381496
Web Page: http://www.kln.gov.my

Embassy of Mali
Zawiat Al Dahmani, Tripoli
Tel: + 218 - 21 - 4839119
Fax: + 218 - 21 - 4839119

Embassy of Malta
Ben Ashour Street, Tripoli
Tel: + 218 - 21 - 3611181
Fax: + 218 - 21 - 3611180

Embassy of Morocco
Ben Ashoor Street, Tripoli
Tel: + 218 - 21 - 3617808
Fax: + 218 - 21 - 3614752

The Royal Netherlands Embassy
Embassy/ The Commercial Section, 20 Jalal Bayar, Tripoli
Tel: + 218 - 21 - 4441549 / 4441550
Fax: + 218 - 21 - 4440386
Contact person: Mr. Samir Yousif

Embassy of Nigeria
Tripoli
Tel: + 218 - 21 - 4443036 / 4443037
Fax: + 218 - 21 - 4443035
Web Page: http://www.nigembtripoli.org

Embassy of Pakistan

Ben Ashour Street, Tripoli
Tel: + 218 - 21 - 3610937
Fax: + 218 - 21 - 3600412

Palestinian Authority

Al Dahra Street, Tripoli
Tel: + 218 - 21 - 4448599
Fax: + 218 - 21 - 3336161

Embassy of the Philippines

Gergaresh Street, Tripoli
Tel: + 218 - 21 - 4833966
Fax: + 218 - 21 - 4836158

Embassy of Poland

Ben Ashour Street, Tripoli
Tel: + 218 - 21 - 3615972
Fax: + 218 - 21 - 3615199

Embassy of Romania

Ben Ashour Street, Tripoli
Tel: + 218 - 21 -3612912
Fax: + 218 - 21 - 3607597

Embassy of the Russian Federation

Madinat Al Hadaeeq, Tripoli
Tel: + 218 - 21 - 3330546
Fax: + 218 - 21 - 3330546

Embassy of Rwanda

Hai Al Andalus, Tripoli
Tel / fax: + 218 - 21 - 4772864 / 4770317

Saudi Arabia Embassy

Al Dahra - Al Qairawan Street, Tripoli
Tel / fax: + 218 - 21 - 3330485 / 4447180

Embassy of Somalia

Gergaresh Gourji, Tripoli
Tel: + 218 - 21 - 4781368
Fax: + 218 - 21 - 3610958

Embassy of Spain

Al Dahra Near Al Qadisya Square, Tripoli
Tel: + 218 - 21 - 3333275 / 3336579
Fax: + 218 - 21 - 4443743

Embassy of the Sudan

Gergaresh, Tripoli
Tel: + 218 - 21 - 4775361
Fax: + 218 - 21 - 4774781

Embassy of Switzerland

Ben Ashour Street, Tripoli
Tel: + 218 - 21 - 3614118
Fax: +218 21 3614328

Embassy of Syria

Al Qadessya Square, Tripoli
Tel / fax: + 218 - 21 - 3331783 / 3339030

Embassy of Togo

Al Dahra, Tripoli
Tel: + 218 - 21 - 4449565
Fax: + 218 - 21 - 3332423

Embassy of Turkey

Zawiat Al Dahmani, Tripoli
Tel: + 218 - 21 - 3401140
Fax: + 218 - 21 - 3401146

Embassy of the United Arab Emirates

Hai Al Andalus - Abu Nawas, Tripoli
Tel: + 218 - 21 - 4832595
Fax: + 218 - 21 - 4832598

Embassy of Vietnam

Gergaresh, Tripoli
Tel / fax: + 218 - 21 - 830674 / 830994

BUSINESS RESOURCES

AUTHORITIES, BOARDS & BUSINESS ASSOCIATIONS

Libyan Foreign Investment Board

Ben Gashir Road n° 20
Tripoli
Tel: +218 21 3618686
Fax: +218 21 3617918
E-mail: rajab@investinlibya.com
Website: www.investinlibya.com

General Board of Ownership Transfer of Companies & Public Economical Units

P.O. Box 5671
Tripoli
Tel: +218 21 4891540
Fax: +218 21 4890015
Website: www.tamleek.gov.ly

General Union of Libyan Chambers of Commerce, Industry & Agriculture

26 Bandong st. P.O. Box 12556
Tripoli
Tel: +218 21 4442821, 4441457
Fax: +218 21 3340155
E-mail: unionchambers@hotmail.com

Libyan Businessmen Council

Dat El-Imad Tower 5, 1st floor, P.O. Box 91491
Tripoli
Tel: +218 21 3350213/4
Fax: +218 21 3350374

Tourism Investment & Promotion Board

P.O. Box 91871
Tripoli
Tel: +218 21 3405112/3
Fax: +218 21 3405115
E-mail: tipblibya@yahoo.com

Companies & Commercial Registration Department

Tripoli
Tel: +218 21 4808704
Fax: +218 21 4808705

United Nations Development Programme

69/71 Turkia St. P.O. Box 358
Tripoli
Tel: +218 21 3330852
Fax: +218 21 3337349

General Board of Information & Documentation

P.O. Box 12071
Tripoli
Tel: +218 91 3777366
E-mail: nida@nidaly.org

Petroleum Research Center

Tourist City, P.O. Box 6431
Tripoli
Tel: +218 21 4830022/7
Fax: +218 21 4836820
E-mail: admin@prclibya.org
Website: www.prclibya.org

General Syndicate for Petroleum & Mining

P.O. Box 1031
Tripoli
Tel: +218 21 4442076, 4446413
Fax: +218 21 4446568
E-mail: lgupcf_oil@yahoo.com

Oil & Gas Downstream Investment Committee

Dat El-Imad Tower 4, P.O. Box 5335
Tripoli
Tel: +218 21 3350895/6
Fax: +218 21 3350894

CONFERENCE CENTERS & SERVICES

Tripoli International Fair

Omar Muktar Street
Tripoli
Tel: +218 21 4440655, 3332255/9
Fax: +218 21 4448385, 3336175
E-mail: info@tripolifair.org
Website: www.tripolifair.org

Corinthia Bab Africa Hotel

Souk Al Thulatha Al Gadim
Tripoli
Tel: +218 21 3351990
Fax: +218 21 3351992
E-mail: mgauci@corinthia.ly
Website: www.corinthiahotels.com

Golden Star

Tel: +218 21 4440987
Fax: +218 21 4443569
E-mail: goldenstar@mail.com

The Golden Star team working on the organisation of the successful Libya Opportunity and Challenge Event held in Tripoli last year

Alsanawbar Park

Alnaser Jungle.
Tripoli
Tel: +218 21 3609859, 3606797, 3602968
Fax: +218 21 3606765
E-mail: alsanawbar@excite.com

Mohame Zeglam real estate renting

Abousetta rd. behind planetarium and close to Mahari Hotel
Tripoli
Tel: +218 21 3403988
Mobile: +218 91 2158977, 2201932
E-mail: yzeglam@yahoo.com

TRANSLATORS

Magharief Bureau For Legal Translation

Magharief Str.No.215. P.Box: 4467.
Tripoli
Tel: +218 21 3336811, 3331733
Fax: +218 21 3331733
E-mail: magarif@hotmail.com
Website: www.magariftranslation.com

Eddahra Office For Legal Translation

Eddahra Str.-Opposite National General Maritime Transport Company
Tripoli
Tel: +218 21 3341872
Fax: +218 21 4448882

Mohamed Ben Amer Addalil Legal Translation Bureau

Almgharief Str. 229
P.Box: 2379.
Tripoli
Tel: +218 21 3334225, 4448822

Kashadah & Co.

4 Damascus Str.
P.Box: 4769.
Tripoli
Tel: +218 21 3330941, 3334289
Fax: +218 21 4446888
E-mail: akashadah@deloitte.Com

Jamil H.Taher International Office For Legal And Authentication Translation

3 Haiti Street (126)
P.Box: 20328.
Tripoli
Tel: +218 21 3339472
Fax: +218 21 4445914
E-mail: alami2@yahoo.com

Luai M. Ghnegiwa Sworn Translator

El-Baladia Str. No. 81.83
P.Box: 506.
Tripoli
Tel: +218 21 3335244
Fax: +218 21 3335244

Ali Al-Musrati, an experienced and English-speaking car driver in Libya

TRANSPORT SERVICES

Yamama Courier
Tripoli
Tel: +218 21 4440164, 3331962
Fax: +218 21 3336566

ADM Cargo Express Services (TNT)
Gurgi Rad, Hay Al-Andalus P.O. Box 81206
Tripoli
Tel: +218 21 4778008, 4770326
Fax: +218 21 477 8009

E-mail: admlyexpress@yahoo.com
Website: www.tnt.com

African Sail Shipping Co.
El Fateh Tower 2, floor nº 3, P.O. Box 93043
Tel: +218 21 3351387, 3351388
Fax: +218 21 3351390
E-mail: assa2000-ly@lttnet.net
Website: www.assa-agent.com

Overseas Shipping Company Libya
Dat El Imad Tower 5, 1st floor.
Tripoli
Tel: +218 21 3350870-1
Fax: +218 21 3350322
E-mail: oscl@lttnet.net

El Mohandes Company
Elherwi St. P.O. Box 577
Tripoli
Tel: +218 21 3331318
Fax: +218 21 3330084
E-mail: al_mohands2003@yahoo.com

Ali Al-Musrati Taxi Services
Al-Kabir Hotel.
Tripoli
Tel: +218 91 3759144

PRIVATE CONSULTANTS & ACCOUNTANTS

Mohamed Ghattour & Co.

Dar Almansora Building, Jamhuria Street, Almansora District P.O. Box 6026
Tripoli
Tel: +218 21 4444468,
    +218 21 4441004
    +218 21 4447142

One of the offices of Mohamed Ghattour & Co.

Fax: +218 21 3334814
E-mail: ghattour110@yahoo.com
Website: www.pwc.com/me

R.S.Col. Ashur Emgeg Consulting
Saad Zaglug st. n° 12, P.O. Box 4179
Tel: +218 91 2129484
Fax: +218 21 4447966
E-mail: ashur_emgeg41@yahoo.com

Kanoun & Co.
Ahmedi Swehli Street, Off Ennaser Street,
P.O. Box 81924
Tripoli
Tel: +218 21 3338195, 3330976
Fax: +218 21 3336744
E-mail: kanoun41@hotmail.com

Tripoli
Tel: +218-21 33 50 126 /+218-21 33 50 127
Fax: +218-21 33 50 125

Majeri Law Office
P.O.Box 2357 Tripoli Libya
Tripoli
E-mail: info@majeri.com
Website: www.majeri.com/

Tumi Law Firm
190 Khalid Ben Walid St Dahra Area,
P.O.Box: 4444
Tripoli
Tel: +218 21 3332144 /+218 21 3339024
Fax: +218 21 4446097
Website: http://www.tumilawfirm.com

LEGAL REFERENCES

Maghur And Partners Attorneys at Law
20 Khalid Ibn Elwaled Street, Dahra,
P.O. Box 2111
Tripoli
Tel/Fax: +218 21 3331312
Website: www.maghurandpartners.com

Ibrahim Legwell Law Firm
189 First of September Street, P.O.Box 484
Tripoli
Tel: +218-21-3336578 / 3343729
Fax: +218-21-3343730
E-mail: info@legwell.com
Website: www.legwell.com

Mayet Associates Attorneys Counselors at Law
Dat El-Imad Tower 5 Fl. 1, P.O. Box 91197

INSURANCE COMPANIES

Libya Insurance Company
Almgarif st. P.O. box 2438
Tripoli
Tel: +218 21 4444179
Fax: +218 21 4444178

United Insurance Company
Al Tateh Towers, P.O. Box 91809
Tripoli, Libya
Tel: +218 21 3351140/9
Fax: +218 21 3351150/1

Arab Union Contracting Company

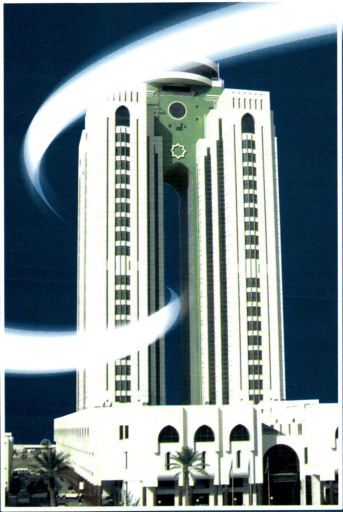

The heart of business in Tripoli

IT, PRINTING & DESIGN SPECIALISTS

MPS Marketing North Africa Ltd.
Ben Ahsour St.
Tripoli
Tel/Fax: +218 21 3609846
E-mail: salem@zhlgraphics.com

ZHL Graphics
Mansoura St.
Tripoli
Tel/Fax: +218 21 3613449, 3618840

Alghad IT Services
Aswae Al Ketoony Str. P.O. Box 91590
Tripoli

Tel: +218 21 3341503
Fax: +218 21 3341503
E-mail: alghad@lttnet.net

Alshafak
El Nakhia St. Dahra
Tripoli
Tel: +218 21 3336431
Fax: +218 21 3330981
E-mail: n.rayes@alshafak.com
Website: www.alshafak.com

HOSPITALS

You can find plenty high standard hospitals in Libya:

In Tajura is also produced a local brand of tractors

Tripoli Medical Centre

University Road
Tripoli
Tel/fax: +218 21 462370 / 710

Central Hospital

Tripoli
Tel: +218 21 3605001 / 10
Fax: +218 21 3339161

Central Heart and Cardiac Medical Center

Tripoli
Tel: +218 21 3692301 / 5
Fax: +218 21 3692301

General Hospital

Health 24 hr Maternity cover
Sourman
Tripoli
Tel: +218 21 3605001 / 10
Fax: +218 21 3624935

Abi Sita Hospital for Respiratory Disease

Tripoli
Tel:+218 21 3508652 / 4
Fax:+218 21 3501075 / 6

Children Hospital

Tripoli
Tel/fax:+218 21 4444181 - 84

Jala Maternity Hospital

Health 24 hr Maternity cover
Omar Mukhtar Street
Tripoli
Tel/fax:+218 21 444-4101

Central Medical Laboratory

Tripoli
Tel:+218 21 3336873
Fax: +218 21 3334414

Cosmetic Surgery Medical Center

Tripoli
Tel: +218 21 605450
Fax: +218 21 3332168

Diabetes Hospital

Tripoli
Tel: +218 21 503068
Fax: +218 21 503055 - 57

Emergency Hospital

Tripoli
Tel/fax:+218 21 4442555 - 57

Kidney Disease and Surgery Medical Center

Al Zahra
Tripoli
Tel/fax:+218 21 4443742

National Centre for Diabetes Endocrinology

Health 24 hr Emergency Room
Beach Road Abusitta
Tripoli
Tel/fax:+218 21 350-1192/3

Tripoli

Al Khadra Hospital
Tripoli, Libya
Tel: +218 61 4904700
Fax: +218 21 4906030

Al Khoums

General Hospital

A busy street in Tripoli near the old Medina

Al Khoums
Tel: +218 31 23689u
Fax: +218 31 23611 - 16

CLINICS

Libyan General Medical Council
P.O.Box 7768
Tripoli
Tel: +218-21-361-4056
Fax: +218-21-361-0356

Al Yousar Clinic
Elsayde Street
Tripoli
Tel: +218 214444799
Fax: +218 91 2121449

Al Afia Clinic
Gasser Benghashir P.O. Box 97521
Tripoli
Tel: + 218 22 633051 / 4
Fax: +218 21 3333055

Al Amal Clinic
Bab Ben Gheshier
Tripoli
Tel/fax: +218 21 608165

Al Masarra Clinic
Ben Ashoor Street
Tripoli
Tel/fax: +218 21 4446481

Rahma Clinic
Tripoli
Tel/fax: +218 21 4623391 - 4623146 - 4623830

Maya Studios

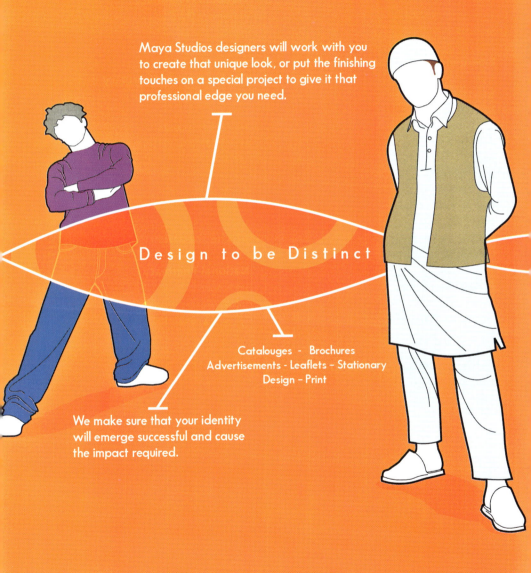

Maya Studios designers will work with you to create that unique look, or put the finishing touches on a special project to give it that professional edge you need.

Design to be Distinct

Catalouges - Brochures
Advertisements - Leaflets – Stationary
Design – Print

We make sure that your identity will emerge successful and cause the impact required.

Tel: +218 21 361 4152   Fax: +218 21 361 4152   info@mayastudioz.com
w w w . m a y a s t u d i o z . c o m

Aboharida Clinic
Algamhoria Street
Tripoli
Tel: +218 21 4442364
Fax: +218 21 4622652

Emergency Ambulance
Tripoli
Tel: +218 213619682 / 84
Fax: +218 21 3619683

Alsheefa clinic
Tripoli
Tel: +218 21 4839024 / 4834775

Dar Al-Shifa Clinic
Znata Street
Tripoli
Tel/fax: +218 21 14624935

Albassatein Clinic
Zawia
Tel: +218 23-621063
Fax: +218 23-629447

Al-Hikma Medical Services co.
Sidi Abdelwahed
Zawia
Tel/fax: +21823629447

PHARMACIES

Tarabulus Pharmacy
Al Sharaa Al Gharbi St.
Tripoli
Tel: +218 21 4802591

Golden Nawras Pharmacy
Opposite of Dat Al-emad
Tripoli
Tel: +218 21 3340624
M: +218 21 (091) 5090443

Central Pharmacy
Mezran St.
Tripoli
Tel: +218 21 3337746

National Pharmacy
Sidi Al Masri St.
Tripoli
Tel: +218 21 3607216

INVESTMENT & LEGAL FRAMEWORK

eBiz
guides

INVESTMENT LEGISLATION

The government aims to open up the Libyan economy to further industrialisation with increased private sector participation and investment of foreign capital. Law No. 5 of 1426 (1997) was enacted to attract the investment of foreign capital. This Law aims at encouraging foreign investment within the framework of the general policy of the State and is targeting the economic and social development, particularly in regard to the transfer of modern technology, the development of Libyan technical resources, the development of national products and their distribution in international markets.

Libya has investment promotion agreements with Egypt, Iraq, Italy, Jordan, Mauritania, Palestine Authority, Somalia, Syria, United Arab Emirates, Yemen.

Below are some of the most interesting articles of Law No. 5 for the year 1426 (1997) concerning the encouragement of foreign capitals investment Issued by the Libyan General People's Congress:

DOMESTIC LAWS: LAW N° 5

Article 1
The aim of this Law is to attract investment of foreign capital in investment projects within the framework of the general policy of the state and for the objectives of economical and social development and in particular:

- Transfer of modern technology.
- Training the Libyan technical workforce.
- Diversification of income resources.
- Contribution to the development of the national products so as to help in their entry into the international markets.
- Realization of a local development.

Article 2
This Law shall apply to the investment of foreign capital held by Libyans and the nationals of Arab and foreign states in investment projects.

It is permitted for local capital to have partnership with foreign capital in investment and the executive regulation of this law shall determine the bases and conditions for the implementation of this partnership.

Article 3
In the application of this Law, unless the context indicates otherwise, the following words and phrases shall have the meaning explained hereunder:

1. Jamahyria means the Great Socialist People's Libyan Arab Jamahyria.
2. The Law means the Law of Foreign Capitals Investment Encouragement.
3. The Secretary means the Secretary of the General People's Committee for Planning, Economy and Commerce.
4. Authority means Libyan Foreign Investment Board.
5. The executive regulation means the regulation issued for the implementation of the provisions of this Law.
6. The foreign capital means the total financial value brought into the Great Jamahyria whether owned by Libyans or foreigners in order to undertake an investment activity.
7. National capital: The monetary or physical value evaluated by local currency included in the formation of the capital of the investment project owned by Libyan citizen or by Libyan juridical personality fully owned by Libyan citizens.
8. Project means any economic enterprise established in accordance with this Law the result of its work is the production of goods for end or intermediate consumption, or investment goods, or for export, or provision of service, or any other enterprise approved as such by the general people's committee.

9. Investor means any natural or judicial entity national or non-national, investing in accordance with the provisions of this Law.

Article 4

This Law regulates the investment of foreign capital brought into the Jamahyria in any of the following forms:

- Convertible foreign currencies or substitutes thereof brought through official banking methods.
- Machinery, equipment, tools, spare parts and the raw materials needed for the investment project.
- Transport means that are not locally available.
- Intangible rights; such as patents, licenses, trade marks and commercial names needed for the investment project or operation thereof.
- Reinvested parts of the profits and returns from the project. The executive regulation shall regulate the manner for the evaluation of the portions used in the formation of the capital designated for investment in the Jamahyria.

Article 5

The authority known as the "Libyan Foreign Investment Board" having its own independent juridical personality, under the jurisdiction of the General People's Committee for Planning, Economy and Commerce is established by a decision from the General People's Committee upon a proposition by the Secretary stating the authority's legal address, its secretary and members of management committee. The executive regulation shall regulate the meetings of the authority and the administrative procedures required for establishing the project.

Article 6

The authority shall encourage foreign capital investment and promotion for the investment projects by various means; in particular it shall:

1. Study and propose plans to organize foreign investment and supervise foreign investments in the country.
2. Receive the applications for foreign capital investments to determine whether they satisfy the legal requirements; and the feasibility study for the project and then submit its recommendations to the secretary accordingly.
3. Gather and publish information and conduct economic studies relevant to the potentials of investments in the projects that contribute to the economic development of the country.
4. Take proper actions to attract foreign capitals and promote the chances of investment through various means.
5. Recommend exemptions, facilities or other benefits for the projects that are considered important for the development of the national economy, or recommend the renewal of the exemptions and benefits as provided for in the Law for further periods of time. It shall submit its recommendations to the relevant authority.
6. Consider without prejudice the right of the investor to petition and litigate complaints, petitions or disputes lodged by the investors resulting from the application of this Law.
7. Study and review periodically the investment legislations, propose improvement thereof and submit same to the concerned authority
8. Perform any other functions assigned to it by the general people's committee.

Article 7

The project must comply with all or some of the following requirements:

- Production of goods for export or contribution to the increase of export of such goods or substitute imports of goods in total or in part.
- Make available positions of employment for Libyan manpower, train and enable some of the workforce to gain technical experience and know-how. the executive regulation shall set the conditions and terms of employment of Libyan manpower.

- Use of modern technology or a trade mark or technical expertise.
- Provision of a service needed by the national economy or contribute to the enhancement or development of such service.
- Strengthen the bonds and integration of the existing economic activities and projects or reduce the cost of production or contribute in making available materials and supplies for their operations.
- Make use or help in making use of local raw materials.
- Contribute to the growth and development of the remote or underdeveloped areas.

Article 8

Investment is encouraged in the following areas:

- Industry
- Health
- Tourism
- Services
- Agriculture
- And any other area determined by a decision from the General People's Committee according to a proposal from the Secretary.

Article 9

The permit for foreign capital investments shall be granted by the authority after the issuance of the secretary's decision approving the investment.

Article 10

Projects established within the framework of this Law shall enjoy the following benefits:

a) An exemption for machinery, tools and equipment required for execution of the project, from all custom duties and taxes, and taxes of the same impact.
b) An exemption for equipment, spare parts and primary materials required for the operation of the project, from all custom duties and custom taxes imposed on

imports as well as other taxes of the same impact for a period of five years.
c) Exemption of the project from the income taxes on its activities for a period of five years as from the date of commencement of production or of work, depending on the nature of the project. This period shall be extendable by an additional duration of three years by a decision from the general people's committee upon a request of the same by the secretary. Profits of the project will enjoy these exemptions if reinvested. The investor shall be entitled to carry the losses of this project within the years of exemption to the subsequent years.
d) Goods directed for export shall be exempted from excise taxes and from the fees and taxes imposed on exports when they are exported.
e) The project shall be exempted from the stamp duty tax imposed on commercial documents and bills used.

Exemptions mentioned in paragraphs a, b and contemplated in this article do not include the fees due for services such as harbour, storage and handling.

Article 11

Equipment, machinery, facilities, spare parts and primary material imported for the purpose of the project may neither be disposed of through sale or abandoned without the approval of the authority and after payment of custom duties and taxes imposed on importation thereof; nor be used for purposes other than those licensed thereof.

Article 12

The investor shall have the right to:

a. Re-export invested capital in the following cases:
　　– End of the project's period.
　　– Liquidation of the project.
　　– Sale of the project in whole or in part.
　　– Elapse of a period of not less than five years as of the issuance of the investment permits.

b. Re-transfer the foreign capital abroad in same form in which it was first brought in after the elapse of a period of six months as of its importation in cases where difficulties or circumstances out of the investor's control prevent its investment.

c. It is permissible to transfer annually the net of the distributed profits realized by the project and interest thereof.

d. The investor has the right to employ foreigners whenever the national substitute is not available:

– The foreign employees who come from abroad have the right to transfer abroad a percentage of their salaries and wages and any other benefits or rewards given to them within the framework of the project.

– Conditions and terms regarding the implementation of this article shall be set by the executive regulation.

Article 13

The investment project shall not be restricted by the legal forms required by the current legislation nor be subject to

The second of the five towers of the Dat-El Imad business complex, where can be found the offices of the Libyan Arab Foreign Bank

the procedures of registration at the commercial registrar nor at the registrar of importers and exporter.

The executive regulation will set the legal forms of investment project wich can be set up according to this law, stablishment conditions and procedure of setting up of such project and registration at the investment registrar. The investment project shall enjoy the status of independent financial and juridical personality once the process of the registration of it's entity is completed

Article 14

A project established in the local development areas or a project which contributes to food security or a project which uses installation and means conducive to save energy or water or contributes to the protection of environment, will enjoy the exemption mentioned in paragraphs b) and c) of article (10) of this Law for an additional period by a decision from the General People's Committee upon a proposal from the Secretary. The executive regulation will set the terms and conditions according to which the project could be considered within these areas.

Article 15

Notwithstanding ownership Laws in force, the investor shall be entitled to hold title for land use. The investor may also lease such land, construct buildings thereon and be entitled to own any property or lease thereof required for establishment or operation of the project; all as per the terms and conditions set in the executive regulation.

Article 16

The investor shall have the right to open for his project an account in convertible currencies at a commercial bank or at the Libyan Arab Foreign Bank.

Article 17

Ownership of the project may be transferred in whole or in part to another investor with the approval of the authority; the new owner will replace its predecessor in all rights, undertakings and obligations arising therefrom, in accordance with the provisions of this Law and other legislations in force. The executive regulation shall set the terms and conditions for the transfer of ownership.

Article 18

In case it is proven that the investor has violated any provisions of this Law or the executive regulation, the authority shall issue a warning to the investor to rectify the violations within a period of time specified therein. In case of failure by the investor to adhere thereto, the Secretary, upon a recommendation by the authority, may:

- Deprive the project from some of the benefits provided for this Law.
- Oblige the investor to pay double the exemptions granted to him.

Article 19

The permit of the project may be withdrawn or the project finally liquidated in the following cases:

- Failure to start or complete the project in accordance with the terms and conditions set by the executive regulation.
- Violation of the general provisions of this Law and executive regulation.
- Repetition of violations.

All in accordance with the procedures specified by the executive regulation.

Article 20

The investor shall be entitled to petition in writing against any decision affecting him as per article (18) or article (19) of this Law, or against any disputes arising because of the implementation of the provisions of this Law within thirty days as of the date of notifying him by a delivery guaranteed letter; the executive regulation shall specify the proper authority to which petitions should be submitted and processes of petition.

Article 21

The investor should:

- Maintain regular books and records for the project.
- Prepare an annual budget and profit and loss account audited by a chartered accountant as per the conditions set forth in the commercial Law.

Article 22

The employees of the authority designated by a decision from the Secretary shall have the power of the judicial officers to control the enforcement of this Law and to unveil and record the violations and refer same to the competent authority; for this purpose the said employees shall be entitled to inspect the projects and check the books and records relevant to their activities.

Article 23

The project may not be nationalized, dispossessed seized, expropriated, received, reserved, frozen, or subjected to actions of the same impact except by force of Law or court decision and against an immediate and just compensation provided that such actions are taken indiscriminately; the compensation will be calculated on the basis of the fair market value of the project in the time of action taken. The value of the compensation in convertible currencies may be transferred within a period not exceeding one year and according to the rate of exchange prevailing at the time of transfer.

Article 24

Any dispute arising between the foreign investor and the state, due to the investor's act or to actions taken by the state, shall be referred to a court having jurisdiction in the Jamahyria except where there is a bilateral agreement between the Jamahyria and the state to which the investor belongs or where a multi-lateral agreements to which the Jamahyria and the state to which the investor belongs are parties that provide for relevant reconciliation or arbitration, or there is a special agreement between the investor and the state containing provisions in regard to an arbitration clause.

Article 25

Foreign investments in existence on the date of issuance of this Law shall enjoy the privileges and exemptions provided for herein.

Article 26

Provisions of this Law shall not apply to foreign capital invested or to be invested in petroleum projects as per the provisions Law number 25 of 1995, as amended.

Article 28

Law number 37 of 1968 regarding investment of foreign capitals in Libya is hereby repealed and so are any other provisions that may contradict the provisions of this Law.

INTERNATIONAL LAWS AND AGREEMENTS

The EU General Affairs Council on 13 September 1999 decided to lift a number of EU sanctions against Libya, which were imposed in 1986. This followed an earlier decision by the EU in April 1999 to suspend EU sanctions adopted in response to Security Council Resolutions 748 and 883 of 1992 and 1993. The EU measures lifted are restrictions on the freedom of Libyan diplomats and consular personnel; and reduction of the staff of diplomatic and consular missions. More recently, on Monday 13th of October 2004, the EU finally ended its 12 years of sanctions and eased an arms embargo to reward Libya for giving up efforts to develop weapons of mass destruction. The U.N. Security Council lifted 11-year-old sanctions in 2003, and in April 2004 the United States removed most of its commercial sanctions.

Libya is a member of the following international organisations:

ABEDA, AFDB, AFESD, AL, AMF, AMU, CAEU, CCC, ECA, FAO, G-77, IAEA, IBRD, ICAO, ICRM, IDA, IDB, IFAD, IFC, IFRCS, ILO, IMF, IMO, Intelsat, Interpol, IOC, ISO, ITU, NAM, OAPEC, OAU, OIC, OPEC, PCA, UN, UNCTAD, UNESCO, UNIDO, UNITAR, UPU, WFTU, WHO, WIPO, WMO, WTO

The Al-Fateh towers are an icon of the business life of Tripoli

IMPORT/EXPORT RULES AND LEGISLATIONS

Trade Policy

Libya seeks to strike a balance between the increased pressures for liberalization of the economy and attracting foreign capital on the one hand and protecting local industry and manufacturing and other national interests on the other.

Imports and Exports

– Imports are generally unrestricted except for specifically banned items.

– Import licences are required.

– Goods of Israeli origin are not permitted to be imported.

– All industrial supplies, foodstuffs and consumer goods are imported into Libya there are few restrictions on imports except for certain prohibited items.

– Imports are subject to customs duties depending upon the category of import.

– There are no official price controls.

– Customs clearance procedures require the appointment of a local agent

– Libya has an annual International Trade Fair where many countries take stalls to exhibit their products.

– Exports are encouraged and are generally free of restrictions.

LABOUR RELATIONS & SOCIAL SECURITY

Manpower

The number of the Libyan manpower according to 2001 was equal to 805580, 67.80% of them were males and 32.20% females. 37.23% of the total manpower is scientific and technical professionals. Another 19.13% is working in the industry, transportation and agriculture fields. Finally, 18.72% are working in services and trade. The secretaries of the popular congresses and people's committees in addition to the directors, executive managers and administrative officers represent as well 18.72% of the total national manpower. Libya is characterized by the availability of cheap and skilled manpower. Moreover the percentage of the economically active category is higher in percentage in young people.

Labour Relations

Availability of Labour

Libya has a work force of approximately one million, of which 31% work in industry, 27% are involved in services, 24% in government and 18% in agriculture.

Employer/Employee Relations/Unions.

The employer/employee relationship is defined in the Libyan Labour Law No. 55 1970 and its subsequent amendments and modifications. The provisions of the Libyan Labour Law 1970 deal with all employer/employee relations such as minimum wage, daily and weekly working hours, night shift regulations, dismissals and training schemes. Labour Law in Libya is fairly comprehensive and specific advice should be sought on the subject. All employees are members of their own employee unions, however the relations between the employers and employees generally tend to be trouble free.

Employee Training Programmes

All employers are expected to have formalised training programmes for the Libyan nationals they employ and are required to demonstrate the implementation of their programmes.

Working Conditions

Working conditions are defined in the Labour Law 1970 and the Social Securities Act 1980.

Wages & Salaries

The levels of salaries and wages payable in the public sector are stipulated by Law 15 of 1981. However the salaries in the private sector vary according to the degree of skills, professional qualifications and experience.

Hours Worked

The normal working week is 42 hours although the Law permits a maximum of up to 48 hours per week. Overtime is paid at the rate of time and a half for week days and double time for Fridays and public holidays.

Paid Holidays & Vacations

The minimum statutory period for annual vacation is 24 working days.

Equal Opportunities

The Libyan Government is committed to promotion of equal opportunities for all its people.

Termination of Employment

A minimum prescribed notice of termination of employment is required under the Law. If it is considered that a particular dismissal is unfair the employee may take the case to court. If the court finds in favour of the employee it may order reinstatement or compensation.

Social Security System INAS

Social Security contributions are due in accordance with Law 13 of 1980 as amended by Law 1 of 1991.

Coverage

The contributions are payable by all employees working in Libya whether local or foreign, based on gross income. Contributions may be made either weekly or monthly. The gross salary with regard to foreign nationals is required to include an amount for housing and subsistence regardless of whether this is paid to the employee or not.

	Foreign companies	Domestic companies
Employee	3.75%	3.75%
Employer	11.25%	10.50%
Government	- 15.00%	0.75% 15.00%

Solidarity Fund

In addition to the above, Solidarity Fund Contributions are payable by deduction from the employees salary at the rate of 1% of gross salary.

Libyan employees are subjected to 1.5% contribution tax (Libyan nationals only)

Foreign Personnel Work Permits

Foreigners who work in Libya require a working permit which must be applied for in advance by the prospective employer. Permission is normally granted provided the skills and qualifications are those that are permitted and required in Libya and are not specifically prohibited by the Labour Office. The approval for a working permit is given usually after one week after the application is received

SETTING UP A BRANCH IN LIBYA

Libyan Law requires that all companies formed in Libya must be Libyan controlled. Foreign companies operating in Libya tend to operate through a branch or a branch of a subsidiary. A foreign company must register its branch with the Secretariat of Economy and Trade and International Co-operation. Details required to effect registration are included at the end of this chapter. Once registration is completed a five year renewable business licence is issued. Law 5 of 1997 exempts foreign companies investing in certain strategic activities from many registration obligations and formalities. An application for the registration of a branch can only be made after a number of procedures have been complied with. These would include:

1. Preparation and submission of various documents and resolutions as detailed at the end of this chapter.

2. Remittance of the Foreign Currency equivalent of LD 70,000 (approximately USD 55,000) to fund the Branch Capital.

After compliance with the above, an application has to be submitted to the appropriate Secretariat. The application together with the supporting documentation would be first checked by the Secretary of the Registration Committee of the Secretariat. The application may well be rejected if the documents are not prepared in the prescribed format. If the application is approved a business licence will be issued. This process may take anything from three to six months. If the application is refused, reasons for the refusal would be given.

Documentation Formats & Preparation

Particular attention needs to be paid in the preparation of the documentation for registration purposes. All documents must be:

- Endorsed by competent authorities in the country issuing the documents and by the Libyan Embassy (Libyan People's Bureau) in that country. They must be original, carbon copies or photocopies are not acceptable unless they bear an original endorsement of the competent authorities.

- Translated into Arabic by a recognised translator. Translations, which originate outside Libya must carry an endorsement of the official authorities and the Libyan Peoples Bureau.

The documents which are required to accompany the application form must be sorted into the groupings as shown by the list of requirements and attached to the application form in the same order. Documents should be filled in order that may be read from right to left and the Arabic translations must be placed on top of the original documents. Both the Arabic and original documents must be notarised and stamped by the Libyan People's Bureau in the country of origin. An application which fails to comply with the rules and procedures is not considered valid.

TAXATION IN LIBYA

The principal taxes in Libya are:
- Revenue Duties
- Corporate Tax
- Salaries and Wages Tax
- Jehad Tax
- Withholding taxes

Taxation issues in Libya are important particularly for foreign investors. Rules are not always interpreted consistently and practices are prone to change with little notice. Local tax advice and assistance is therefore essential.

Revenue Duty

Any contract negotiated in Libya for anything other than a direct supply must be registered with the Tax Department within 60 days of signing the contract. A duty of 2% of the total contract value and 1% on sub-contracts is payable on registration. All invoices must bear the tax department stamp to show that the registration duty has been paid. There is a 5% sales stamp duty on local sales which is added to the sales invoice.

Corporation Tax

Corporation tax is charged under the Law 11 of 2004 previously Law 64 of 1973. This is done in two stages:
1) Preliminary assessment
2) Final assessment

Corporation tax rates range from 15% to 40% of the assessed profit.

Salaries and Wages Tax

Libyan Salaries and Wages tax applies to all salaries, wages, bonuses and benefits which arise from employment in Libya. Tax rates range from 8% to 15%.

Jehad Tax

This tax is payable under Law 44 of 1970 and is levied on personal incomes at 3% and corporation profits at 4%.

Withholding Taxes

Government bodies withhold a stamp duty of 0.5 % from all payments made by them.

A further 0.2 % is also payable on any official receipt including receipts for contract registration duties, corporation tax, personal tax etc.

Other Tax Notes

- There is no gift and inheritance tax
- There is no VAT
- There are no local government taxes
- A company in receipt of an invoice from a contractor should always ensure that the invoice has been registered with the Tax Department. The company could become liable to the registration tax if this is not paid. The company should further ensure that the contractor is registered with the Tax Department and should sign the tax certificate periodically
- The export of goods to Libya is not subject to tax in Libya if the supplier's commitments end before customs clearance and the supplier is not registered in Libya. If the supply of goods is accompanied by a contract for installation or commissioning that contract will be taxable.

Double tax agreements

- Libya has double taxation agreements with Algeria, Egypt, Iraq, Italy, Jordan, Malta, Mauritania, Morocco, Palestine, Somalia, Sudan, Syria, Tunisia, United Arab Emirates and Yemen
- There are double taxation agreements on civil aviation activity with Italy, Kingdom of Saudi Arabia, Kuwait, Malta and Poland

- There are agreements in respect of INAS (social security) with Algeria, Bulgaria, Greece, Malta, Mauritania, Morocco, Pakistan, Poland, Romania, Tunisia, Turkey and Yugoslavia.

INVESTMENT OPPORTUNITY

The Libyan economy depends primarily on revenues from the oil sector, meaning practically all export earnings and about one-quarter of GDP come from this sector. These oil revenues and a small population give Libya one of the highest per capita GDPs in Africa. Libya in the past three years has made progress on economic reforms as part of a broader campaign to reintegrate into the international fold. This effort picked up steam after UN sanctions were lifted in September 2003. Libya is ahead in liberalizing the economy, initial steps - including applying for WTO membership, reducing some subsidies, and announcing plans for privatisation - are laying the groundwork for a transition to a more market-based economy.

The yearly output of petroleum in Libya exceeds 500 million barrels per year, while the amount of natural gas reaches 10.3 billion m^3?. Libya has a large production of refined products and construction materials.

The non-oil manufacturing and construction sectors, which account for about 20% of GDP, have expanded from processing mostly agricultural products to include the production of petrochemicals, iron, steel, and aluminium. The more popular craft items are carpets, pottery, leather goods, fabrics, and copperware.

The main partners are Italy, making up about 40% of the export market and 18% of the imports, Germany with 20% and 12%, Britain with 6,5% and 3%.

Arable land in Libya is estimated at more than 2 million hectares, in addition to over 13 million hectares of pasturelands.

Agriculture represents about 5% of the total GDP and is mainly producing for the domestic market, but as much as 15-20% of the population is engaged in this sector. Output includes wheat, barley, olives and dates.

Regarding livestock, sheep dominate, counting about 5.6 million. There are about 1.3 million goats, 140,000 cattle, 160,000 camels and 60 million poultry. The most important region for agriculture in Libya is in Tripolitania, but with the construction of the Great Man-Made River, eastern provinces are projected to have increased agricultural output in the years to come.

With a coastal line of 1800 km of clean pollution-free seawater, experience and marine research have indicated ample quantities of white fish and tuna. Local marine environment is naturally suitable for aquaculture investment projects. The unexploited sea sponge and coral reserves represent other interesting areas for marine investments.

A little-known - or poorly-known - heritage, preserved partly by the absence of tourism and offering an array of treasures, has observed the rise and fall of brilliant and sophisticated civilizations on this vast territory. Dating from Prehistory to Islamic Civilization, it is a clear illustration that Libya has a heritage whose incalculable value belongs to all humanity.

The lifting of remaining U.S. and European economic sanctions in September 2004 is another sign of broader investment opportunities.

Foreign Companies Working In Libya

There are over 450 foreign companies' branches registered, with a diversity of nationalities: French (55), Italians (52), British (44), Germans (40), Maltese (18), Swiss (14), Austrians (9), Greeks (6), and Spanish (4). Americans like Panama (6), U.S.A (1) and many other multinationals. They operate in Libya in different sectors, especially in construction, water, electricity and oil field services.

Activities Allowed for Foreign Company's Branches:

Contractual and civil works
– Roads and bridges.
– Dams.
– Maritime embankments and docks and storages.
– Harbours deepening.
– Airports.
– Railways and stations.

Electricity
– Design and implementation of stations of low and high voltage.
– Cable installations.

Oil & Gas
– Exploration and survey of land.
– Examination and analysis of data and geological studies.
– Drilling and maintenance of wells.
– Mud servicing.
– Cementation works.
– Design, installation and maintenance of oil and gas pipes.
– Installation of tanks and pumping stations.
– Installation of sea platforms.
– Removal of mines.
– Installation and maintenance of drilling equipment and plunge pumps.
– Installation and maintenance of safety systems.

Communications
– Design and installation of cordless systems.
– Installation of stations and towers for air navigation systems.

INVESTMENT & LEGAL FRAMEWORK

Industry
– Electrical and hydro mechanical works for plant installation and maintenance.
– Mining for all kinds of minerals and its extraction except for oil.
– Installation and maintenance of industrial ovens.
– Land survey and urban planning
– Seismic and air-turning methods.
– Engineering consultancy for urban and rural areas.
– Environment Protection
– Installation and maintenance of environmental stations and desalination plants.
– Waste recycling.
– Sewage treatment.

IT
– Installation and maintenance of IT systems and programs.

Technical services
– Consulting detailed engineering work.
– Project design.

Health care
– Installation and maintenance of medical equipment.

Foreign Investment

Law N° 7/2003 amended Law N° 5/1997, allowing local capital to participate with foreign capital investors in the investment projects; also new executive regulations were issued. Up to mid 2004, the number of projects that received permits was 102, with accounted total foreign capital participation of over 2000 million USD.

The Tripoli Port is usually very busy with ships coming from many different international destinations

INVESTMENT & LEGAL FRAMEWORK

Areas of Investment

Investment is being targeted in the following areas:

Industry

- Red stone.
- Wall and floor tiles.
- Sanitary equipment.
- Medical injections.
- Extrusion and anodizing of aluminium sections.
- White cement.
- Refrigerators and Air conditions.
- Geysers and Washing machines.
- Chicken hatching.
- Electric heaters and Lamps.
- Grinding used tires and re-production.
- Stupefaction gas.
- Televisions.
- Leather shoes.
- Carpets.
- Fodder (animals, poultry, Fish).
- Air compressors and spare parts.
- Producing the silicate sodium.
- Engines wire wrapped.
- Wool processing.
- Refills paper project.
- Transport containers.
- Producing the concentrating milk.
- Compressed wood

Health

- Build hospitals and clinics.
- Build specialist centres (X-ray / labs).
- Medical equipments.
- Medicine industry.

Tourism

- Resorts and campsites.
- Hotels and villages.
- Desert campsites.
- Treatment centres.
- Tourist training centres.
- Tourist Villages.

Services

- Hospital management.
- Hotel management.
- Harbour management.
- Airports management.

Civil aviation

- Airports maintenance and equipping.
- Airports catering.
- Air transport.

Education

- Universities and high institutions.
- Technical training centres.
- Postgraduate.

THE LYBIAN ECONOMY

OVERVIEW

The Libyan economy depends primarily upon oil sector revenues. These account for practically all export earnings and approximately one-third of the GDP. Libya underwent strong economic growth during 2003, with real gross domestic product (GDP) estimated to have increased by 2.7%-3.8%, from 0.2%-1.5% growth in 2002. For 2004, the anticipated GDP growth is in the region of 2.3%-2.6% with consumer price inflation of 1.9%-3.5%.

The growth of GDP experienced a delay in the eighties, which directly affected all production sector activities and their policies in general, especially the industrial sector. This delay was caused by the repeated decline in oil prices in the late seventies. The participation of the oil sector in the economy was equal to 63.1% and 61.8% of GDP in current prices through 1970 and 1980 respectively. That participation began to fall to approximately 28.3% in 1986, and then went back up to 35.6% in 1992, down to 22.1% in 1998, and to rise to 37.8% in 2000. (See table below).

Table: Participation value of Oil sector & Non-oil sectors, Million LD Current Prices (1970-2000) – Selected years-

Year	1970		1976		1986	
Sector	Value	%	Value	%	Value	%
Oil sector	812.6	63.1	2750.0	57.7	2480.4	28.3
Other sectors	475.7	36.9	2018.1	42.3	6294.0	71.7
GDP	1288.3	100.0	4768.1	100.0	8774.4	100.0

Table continued

Year	1992		1998*		2000*	
Sector	Value	%	Value	%	Value	%
Oil sector	2406.2	35.6	2786.0	22.1	6661.0	37.8
Other sectors	4361.3	64.4	9824.6	77.9	10959.2	62.2
GDP	6767.5	100.0	12610.2	100.0	17620.2	100.0

Source: National Report of Human Development – Libya – 1999, p 80.

* Source: NIDA, Statistical Book 2002.

From the previous table, it is evident that the participation of the oil sector in GDP was reduced while it seems the other sectors' participation had increased. The real cause is not that, but the declined oil prices and its domination upon the national economy, despite all of the huge development efforts taken to reduce the oil sector's participation and to increase that of other sectors, particularly agriculture and industry.

According to some studies, an increase or decline in oil prices by 10% will result in an increase or decline in the national income by 3%. The effect of this relationship reaches the realisable production levels, the supply of commodities and importations, the exchange rates and as a result in price levels, inflation and recession in the national economy.

There were many factors and determinants that came together to raise and secure the status of fluctuation and recession. The most important one is the low growth of oil revenues in times where shortage of hard currency was evident, which in turn resulted in shortages of the necessities for production activities. The other factor is the negative consequences coming from the air embargo in the early nineties and the economic sanctions that were imposed on Libya in the mid-eighties. Both caused severe harm to the economic activities of different sectors. (See tables below).

Table: Losses caused by the air embargo and economic sanctions (1992-1998)

Sectors	Value of Material Losses US$ (000)
Health & Social security	294,000
Agriculture	472,155
Animal wealth	7,187,000
Communication & transport	3,485,000
Industry & Minerals	7,200,000
Economy & Trade	8,200,000
Energy (Oil & Electricity)	7,000,000
Total losses	**33,838,155**

Source: National Report of Human Development – Libya – 1999, p 81.

Figure: Losses caused to Sectors

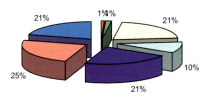

- ■ Health & Social security ■ Agriculture □ Animal wealth
- □ Communication & transport ■ Industry & Minerals ■ Economy & Trade
- ■ Energy (Oil & Electricity)

The country's current economic strategy is to: 1) diversify into non-oil sectors with special emphasis on agriculture, industry and technology, 2) to promote trade agreements with other African nations, 3) move towards a range of economic reforms and privatization and 4) to promote and encourage joint ventures with foreign companies.

Libya is eager to reduce its dependency on oil as the country's only source of income, and to increase investment in agriculture, tourism, fisheries, mining, and natural gas. Being an oil producer, it is especially vulnerable to the fluctuations in international oil prices. Import restrictions have often resulted in shortages of basic foodstuffs.

The agriculture sector is a top government priority. It is anticipated that the Great Man Made River (GMR), a five-phase, US$30 billion project to bring water from aquifers under the Sahara desert to the Mediterranean coast, will alleviate the country's water shortage problem and provide water for agriculture so that domestic agricultural products can reduce its dependence on imported foodstuffs.

Libya is also positioning itself as a key economic intermediary between Europe and Africa. To this end it has become more involved in the Euro-African arena. Libya has also promoted and pressed for a new, regional economic union, called the African Union. The proposed union would ultimately result in an integration of national economies as well as common monetary and fiscal policies. The short term goal of the union would be to encourage investment in sectors such as agriculture and fishing.

Privatization

The economy has been undergoing a gradual process of liberalisation by the government since 1985, but in specific terms (transition of state-owned enterprises to the private sector), it started in 1989. There are laws and decisions concerning privatization, which come in the form of the issuance of regulations for the privatization of certain government-owned enterprises and by allowing private businesses to operate in the country.

Many state-owned enterprises are listed and are under assessment and revaluation for privatization. There is the General Authority for Privatizing the State-Owned Enterprises (according to GPC decision No. 184 of 2001), which contracted some consulting companies for this scope. Foreign participation to invest in the privatized enterprises has already been applied for and considered, side by side with the national participation.

In late July 2003, a conference was organized in Tripoli called "The First National Conference on Privatizing the Public Sector". Local experts and foreigners, participated especially from Britain, Malta and Egypt.

In late 2003, the Secretariat of General Peoples' Committee (SGPC) Dr. Shukri Ghanim[1] declared in a news conference the 5-year privatization program (restructuring, revaluating and transaction to take place 2004-2008). In this program, Libya's economic activities will be transferred and run completely by the private sector and that 360 state-owned enterprises will be privatized. Privatizing processes are already taking place; nationals and some foreign entrepreneurs have bought considerable shares. The privatization program is divided into three phases as in the following table.

[1] *Prime Minister since June 2003. Ex Minister of Trade & Economy.*

Table: Privatization program phases 2004-2008

Activity	Phase 1	Phase 2	Phase 3	Total
Industrial	145	41	18	204
Agricultural	28	4	24	56
Animal breeding	71	-	11	82
Marine	16	1	1	18
Grand Total	260	46	54	360

Phase 1: 01/01/2004-31/12/2005
Phase 2: 01/07/2004-30/06/2007
Phase 3: 01/01/2004-31/12/2008

Participation in WTO

In 2001, Libya applied for membership to the World Trade Organisation without gaining much attention. On Feb 24-25, 2003 a workshop was held in Tripoli on the Libyan participation in WTO and the intellectual property rights, with the participation of the Secretariat of General People's Committee of Economics & Trade and the UN Resident Representative.

In June 2003, the Leader called for privatization of the country's oil sector, in addition to other areas of the economy and pledged to bring Libya into the World Trade Organization (WTO). The former Minister of Trade and Economy, a proponent of privatization, was appointed as Prime Minister. In July 2004, the Libyan application was accepted by WTO and, negotiations have been started to meet the requirements.

Joint Ventures & Foreign Investment

During 2002, a multi-tiered exchange rate system was unified effectively by devaluing the country's currency. Among other goals, the devaluation aimed to increase the competitiveness of Libyan firms and to help attract foreign investment into the country. During October 2003, the Prime Minister announced a list of 360 firms in a variety of sectors including steel, petrochemicals, cement and agriculture to be privatized in 2004.

The Libyan government has established some joint venture companies with foreign participation, for example in the utilities sectors. A number of national corporations have been created in many sectors and, in a growing number of cases, private capital is also participating. Libya has trade and economic co-operation agreements with most European and African countries and also participates in all regional Arab agreements.

Libyan investments abroad under bilateral agreements cover joint companies involved in cattle, agricultural production and exploitation of minerals in Africa, construction in Malta, fishing in Tunisia, and industrial products in Europe and Latin America.

Inflation

The Libyan economy in the last decade, due to its low productivity and the reliance upon one-export commodity of which the international markets govern prices, faced inflationary pressures which resulted in price rises of many commodities and services. The supply of local and imported commodities declined, black markets became evident, public debt increased, issued money in circulation increased in addition to other negative factors. Together, these factors made the inflation rate rise with respect to the consumer prices index from 11.7% in 1991 to about 17% in 1998.

During the period 1999-2002, due to the flexible monetary policy of the Central Bank with respect to the exchange rate and the integration of the commercial policy with the new exchange rate, the control of inflation was made possible. There were decreases in prices of many commodities and services and the Libyan dinar was unified in 2002. The prices decreased by 2.9% in the year 2000 with respect to 1999, and continued decreasing by 9.2% in 2001 with respect to

2002.

In 2003 and 2004 the inflation rate was (-2.1) and (-2.3) respectively.

Assessment of the Economy during 2001

The economy during 2001 had experienced many local and international changes that directly affected the growth rate and the standard of living, the macro and sectarian factors. The main changes were:

a) The introduction of some policies and actions aimed at higher productivity; more work posts and the provision of houses to lower-income levels.

b) The increase of oil prices with respect to the decline of its prices in 1999.

c) The increase of oil income by 1.5% in 2001 with respect to 2000.

d) The decrease of the exchange rate of the Libyan dinar by 11.7%; the introduction of a special exchange rate to reach the dinar balance price and the emphasis upon a flexible monetary policy made it possible to control the inflation and settle prices.

e) The reconstructing of some public production and service projects and companies to enhance its status and efficiency.

Preliminary figures indicate an increase in GDP, from LD 17,620.3 million in 2000 to about LD 18,112.0 million in 2001 based on fixed prices of 2000. That means an increase of LD 491.7 million, with a real growth rate of 2.8%. The non-oil product is expected to grow by 3.3%, due to the increase in the non-oil economic activities, from LD 10,959.2 million in 2000 to about 11,317.5 million in 2001.

As a result of this increase in the non-oil product, an increase in the real personal income is expected with respect to non-oil products from LD 2,086 million in 1999 to about LD 2,138 million in 2000, and a little more than LD 2,147 million in 2001, indicating a compound growth rate of 1.5% during 1999-2001, as the following table indicates.

Table: Growth of GDP and personal income 1999-2001,

Sector/Year	1999	2000	2001	Growth Rate %
Oil sector	6515.3	6,661.1	6,794.5	3.5
Other sectors	10,386.9	10,959.2	11,317.5	4.4
GDP	16,902.2	17,620.3	18,112.0	3.5
Population (Thousand)	4,980.0	5,125.0	5,270.0	2.9
Average Income/Capita LD From non-oil product	2,086.0	2,138.0	2,147.0	1.5

Source: General Board of planning Nov 2002. Annual Report.

In the same year, as a result of the fiscal and banking settlements undertaken by the Secretariat of Treasury and the Central Bank, the value of the public debt decreased to LD 7,644 million, indicating with regard to GDP 42.2%. The following figure illustrates the GDP trend:

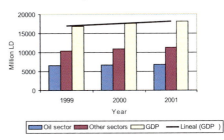

Figure: GDP Trend 1999-2001

Currency

Libyan dinar (LD) = 1000 Dirhams. Notes are in denominations of LD 20, 10, 5, 1, half and quarter.

Coins are in denominations of half, quarter, 100, 50, 20, 10, 5, 1 Dirhams.

Exchange rates

The following figures are included as a guide to the movement of the Libyan dinar (Inter-Bank) against the US dollar and euro:

Date	Jan 01	Jan 02	Jan 03	Jan 04	Jun 04
$1.00 =	0.54	1.30	1.22	1.30	1.31
E1.00 =	0.51	1.18	1.26	1.62	1.60

Interest Rates

Commercial banks' annual interest rate on loans is at around 6%. Foreing companies may obtain loans from Libyans banks provided that they provide a guarantee from their respective home banks. Loans are in LD. This process is not immediate; it may take between 5-6 months. Special interest rates are set for some local loan cases, such as industrial, agriculture and housing projects at 3.0%. The interest rate on deposits is between 3.0% and 5.0%.

Imports and exports

Imports are generally unrestricted except for specifically banned items, and no import licences are required. Exports are encouraged and are generally free of restrictions.

During the year 2003, the value of exports and imports of Libya indicated a trade surplus of LD 9,208.8 million against a corresponding surplus of LD 4,591.3 million in 2002. The value of exports during 2003 increased by LD 4,629.6 million compared to 2002. Whereas the value of imports during the same period increased by LD 12.2 million as compared to 2002. In terms of percentage changes, the increase was 45.5% in exports and, was 0.2% in imports as compared to 2002.

The following table shows the amount of trade balance during the years from 1993 to 2003 and, the annual percentage changes of the trade balance over preceding years.

Table: Balance of trade with actual and percentage changes during 1993-2003

Year	Trade Balance LD (000)	Changes in Trade Balance over the previous year	
		Actual LD (000)	Percent (%)
1993	766,272	(-) 850,475	(-) 52.6
1994	1,629,261	862,989	112.6
1995	1,493,592	(-) 135,669	(-) 8.3
1996	1,663,908	170,316	11.4
1997	1,316,938	(-) 346,970	(-) 20.9
1998	170,293	(-) 1,146,645	(-) 87.1
1999	1,753,594	1,583,301	929.7
2000	3,310,059	1,556,465	88.8
2001	2,733,554	(-) 576,505	(-) 17.4
2002	4,591,326	1,857,772	68.0
2003	9,208,750	4,617,424	100.6

Source: NIDA, External Trade Statistics 2003.

ENERGY

INTRODUCTION

Libya started its oil exploration in 1955 and the principle legislation, Petroleum Law No. 25, was enacted in April 1955. The first oil fields were discovered in 1959 at Amal and Zilten, together known as Nasser, and oil exports began in 1961. Libya became a member of OPEC in 1962. At the beginning of 1999, Libya's OPEC production quota was 1,227 mmbpd.

Now, Libya is Africa's major oil producer and one of Europe's biggest North African oil suppliers. Supplies from North Africa to European destinations have the advantage of being both timely and cost effective.

Libya's economy is based on oil and its export makes up between 75% and 90% of state revenues. Libya has proven reserves of 29.5 billion barrels of oil and a production capacity of 1.4 million barrels per day. During 2003, the oil production was estimated at 1.5 million barrels per day (bbl/d), with domestic consumption of 227,000 bbl/d and net exports of around 1.25 million bbl/d. Libya exports almost 90% of its production to European countries. In 2003, the exports were as follows:

- Italy 485,000 bbl/d
- Germany 188,000 bbl/d
- France 47,000 bbl/d

Spain and Greece also imported oil from Libya.

Libyan oil is mostly sold on a term basis, including to the country's Oilinvest marketing network in Europe, to companies such as Agip, OMV, Repsol YPF, Tupras, CEPSA, and Total. Small volumes are sold to Asian and South African companies. Libyan oil is priced off of Dated Brent.

Product Description

Libya crude is high quality, low-sulphur crude oil produced at a very low cost (as low as $1 per barrel at some fields).

Main export grades of the Libyan crude include Es Sider (36-37o API), El Sharara (44o API), Zueitina (42o API), Bu Attifel (41o API), Brega (40o API), Sirtica (40o API), Sarir (38o API), Amna (36o API), and El Bouri (26o API).

The Industry

Libya has excellent potential for more oil discoveries. Only around 25% of Libya's area is covered by agreements with oil companies. The under-exploration has resulted largely from the sanctions and stringent fiscal terms imposed by Libya on foreign oil companies.

National Oil Corporation (NOC)

Libya's oil industry is managed by the state-owned National Oil Corporation (NOC), a holding company for its subsidiaries, accounting for half of the country's oil output. A number of international oil companies have exploration and production sharing agreements with the NOC.

The NOC's subsidiaries include:

– Waha Oil Company (WOC), formerly Oasis Oil Co., a joint venture of the NOC, Conoco, Marathon, and Amarada Hess; Arabian Gulf Oil Company (Agoco), with production coming mainly from the Sarir, Nafoora/Augila, and Messla fields; Zueitina Oil Company (ZOC), which operates formerly Occidental Inc and Sirte Oil Company (SOC), originally created in 1981 to take over Exxon's holdings in Libya.

Production: Up Stream & Down Stream

Since 1968, the NOC together with its 33 subsidiaries has controlled the entire gas and oil industry, both upstream and downstream. The NOC and its subsidiaries account for 63% of Libya's production. The main subsidiary production companies are Arabian Gulf Oil Company (Agoco), Waha Oil Company (WOC) and Sirte Oil Company (SOC).

Libya's territory is mostly unexplored, offering huge opportunities for the oil industry

Since 1979, the NOC has been allowed to enter into agreements with foreign oil companies. Numerous international companies are engaged in exploration/production sharing agreements with the NOC, the largest being Agip-ENI. Oilinvest is the international arm of the NOC. With its subsidiaries Gatoil and Tamoil, it controls a network of overseas refineries and manages all international investments. UMM Jawwaby Oil Services is the procurement arm for the NOC based in London.

Libya's upstream oil industry is the key to its economy. It is expected to earn US$11.7 billion from oil exports in 2000, which is more than double its 1998 earnings. Oil represents more than 95% of total export revenue and 98% of hard currency earnings. Libya produces high quality, low sulphur crude oil that is highly valued. Its proven reserves are 29.5 billion barrels and production is 1.4 million bpd. This represents less than half of its peak production output of 3.3 million bpd in 1970, a decrease due mainly to the direct and indirect effects of sanctions. Libya would like to increase production

and wants to attract foreign investment to the oil and gas industry.

As in the upstream sector, the NOC controls the whole of the downstream sector together with its numerous subsidiaries and overseas arms, Umm Jawwaby Oil Services and Oilinvest with its two subsidiaries of Gatoil and Tamoil.

The downstream sector was very badly hit by the sanctions and this constrained Libya's ability to increase its supply of products to Europe.

In addition to its oil industry, Libya has an active chemicals industry and is one of the larger markets in the African lubricants industry.

On-Shore Oil

Libya's production is mainly on-shore in three geological areas of the Sirte Basin:

Aerial view of the impressive ENI platform on the Libyan coast

1. The western fairway, which includes several large oil fields namely Samah, Beida, Raguba, Dahra-Hofra, and Bahi.

2. The central northern part of the country, which includes Defa-Waha and Nasser fields, as well as the large Hateiba gas field.

3. An easterly trend that has enormous fields like Sarir, Messla, Gialo, Bu Attifel, Intisar, Nafoora-Augila, and Amal.

Of Libya's existing on-shore oil fields, 12 have reserves of 1 billion barrels or more, and two have reserves of 500 million to 1 billion barrels. Most oil fields in Libya have lives of about 33 years and, with the exception of Murzuq, most of the oil fields were discovered between 1956 and 1972. The NOC's fields are undergoing a natural decline at a rate of 7%-8% per year.

The priority for on-shore exploration includes new areas in the Sirte, Ghadames and Murzuq Basins and in unexplored areas such as Kufra and Cyrenaica. Existing oil field life could also be greatly extended by the application of enhanced oil recovery techniques. Libya faces the challenge of maintaining production at its mature fields such as Brega and Sarir, Waha and Zuetina and bringing new fields such as Murzuk-El Sharara and Mabruk on line.

Off-Shore Oil

There is a relatively narrow continental shelf and slope in the Mediterranean and the Gulf of Sirte. The largest off-shore field is El Bouri, which has proven reserves of 2 billion barrels and a possible 5 billion barrels of oil, and 2.5 Tcf of gas. This field, discovered by Agip-ENI in 1976 is central to Libya's plans.

Eni of Italy is the developer of the El-Bouri oil field off Libya's western coast. This is the largest producing oilfield in the Mediterranean Sea at 60,000 bbl/d approximately. The first phase of field development was completed in 1990 at a cost of US$2 billion. In 1995 El-Bouri was producing about 150,000 bbl/d but this declined due mainly to UN sanctions on import of Enhanced Oil Recovery (EOR) equipment. El-Bouri also has large volumes of natural gas. The El Bouri field was purchased by Repsol YPF in 1993 for US$65 million.

There have been a number of other oil discoveries at various Libyan blocks. El Sharara in the Murzuq basin being a major one currently producing 170,000 bbl/d. Repsol YPF leads a European consortium at the field, consisting of OMV and Total. Oil from El Sharara is processed by the Az Zawiya refinery.

The Al-Jurf offshore oilfield in Block 137 started production in August 2003. Output at Al Jurf is expected to reach 40,000 bbl/d. The NOC holds a 50% share in the field, along with Total (37.5%) and Germany's Wintershall (12.5%).

The petro-chemical industry, which depends on the oil sector for raw materials, has grown rapidly, with large-scale industrial complexes situated at Ras Lanuf and Bu Kammash.

Refineries

The total crude handling capacity of Libya's four refineries is in excess of 350,000 barrels per day. The refineries are all operated by subsidiary refining companies and owned by the NOC. The four refineries are at Brega, Zawia, Ras Lanuf and Tobruk.

There are plans for a new refinery at Sebha. This refinery will be designed to process 20,000 bpd of crude oil from the first phase of the Murzuk field. The anticipated output from Sebha will be 268,000 tonnes/year of gas oil, 184,000 tonnes/year of gasoline, 84,000 tonnes/year of jet fuel and 274,000 tonnes/year of fuel oil. The last will be used in a new power station that is also planned for this area. An export refinery with a capacity of 200,000 bpd is planned for Misurata.

The Brega refinery at Mersa El Brega is a topping and reforming refinery with a capacity of 420,000 tonnes per year (8,400 bpd). The refinery is owned by the Libyan government and is operated by the Sirte Oil Company.

The Ras Lanuf refinery is a topping and reforming refinery with a capacity of 11 million tonnes per year (220,000 bpd) and became fully operational in 1985. The refinery is owned by the Libyan government and is operated by the Ras Lanuf Oil & Gas Processing Company, a subsidiary of the NOC. The refinery produces fuel oil, gas oil, naphtha and kerosene. Expansions to the refinery, which are underway, will permit the production of benzene, butadiene, and MTBE.

Total crude oil production from all Libyan refineries for June 2003 was 1,430,000 bpd (OPEC). Libya aims to boost its oil output capacity by 175,000 bbl/d in 2004 with the help of European companies. The suspension of UN sanctions, along with possible changes to Libya's 1955 hydrocarbons legislation, could be helpful in this regard.

In addition to its domestic refineries, Libya is a direct producer and distributor of refined products in Italy, Germany, Switzerland, and Egypt. In Italy, Tamoil Italia, based in Milan, controls about 5% of the country's retail market for oil products and lubricants; which are distributed through nearly 2,100 Tamoil service stations.

MAIN SECTORS OF THE ECONOMY: ENERGY

Developments

In August 2002, the NOC approved development in the Field A in Block NC-186 in the Murzuq basins. The NOC's priorities for new exploration include areas in Blocks 25 and 36 in the Sirte, Block 20 in Ghadames, plus other unexplored areas such as Kufra and Cyrenaica. Furthermore, the NOC plans to employ modern Enhanced Oil Recovery (EOR) techniques to existing oil fields.

Libya needs to maintain the production at mature fields such as Brega, Sarir, Sirtica, Waha and Zuetina and simultaneously bring into production new fields like El Sharara, with reserves of 2 billion barrels. Repsol YPF, along with Austria's OMV and Total, commenced production of 170,000 bbl/d in December 1996.

An international consortium led by British company Lasmo, including Eni and a group of five South Korean companies, announced that it had discovered recoverable crude reserves in the order of 700 million barrels at the NC-174 Block, 465 miles south of Tripoli, also in Murzuq. Lasmo, which was purchased by Eni in 2001, estimated that production from the field, called Elephant, would cost around $1 per barrel. Elephant originally was due to begin production in 2000 at 50,000 bbl/d, and utilise an existing 30-inch pipeline located 42 miles to the north. Production was delayed and it is now expected to start up in 2004, reaching full capacity of 150,000 bbl/d by 2006.

Its three refineries have a nameplate capacity of 348,000 bpd, which is nearly twice its domestic consumption. The refineries, however, are outdated and desperately in need of upgrading, a matter which has been difficult as sanctions have made equipment and technology less accessible. Libya plans to upgrade its existing refineries and build new refineries.

Foreign Investment & Sanctions

The licensing authority is the Secretariat of Petroleum Affairs. Since 1973, petroleum rights have been granted under a series of production sharing agreements. Decision number 10 of 1979 allowed the NOC to enter into agreements with foreign companies. There have been three exploration and production sharing agreements issued. EPSA-111 remains the model contract in use at the end of 1999. Libya is considering changing the 40-year hydrocarbon legislation to improve terms for foreign investment. The amendments that they are considering will include: access to exploration acreage; small field development; large field incremental production opportunities; increased transparency; and adoption of international competitive bidding practices. In November 1999, in the latest bidding round the acreage was offered under conditions of EPSA-111.

Foreign involvement in Libya was severely reduced as a result of the sanctions and embargoes emplaced upon it, especially between the years of 1992 and 1999. Access to oil industry equipment and technology was restricted and Libya is reliant on foreign investment to keep the industry active. After almost 10 years, sanctions were lifted against Libya in 1999. With the suspension of sanctions, oil companies have shown an eagerness to invest in Libya, and a poll of 76 global oil companies (New Ventures 2000 survey) indicated that Libya is the number one preferred location for oil exploration and production.

The reasons for this are numerous. Libya is Africa's biggest oil producer and Europe's biggest North African oil supplier. It provides extremely high grade, sweet crude. It has very low production costs and the oilfields are close to the refineries and markets of Europe.

Libya welcomes foreign companies to participate in increasing the national oil production capacity from 1.4 million bbl/d at present, to 2 million bbl/d by 2010. To upgrade its oil infrastructure in general and meet the production target, Libya is seeking foreign investment of US$30 billion through 2010. Libya is a highly attractive oil producer given its low cost per barrel and its proximity to European markets, and its well-developed infrastructure.

Recent Joint Venture

The November Seventh concession, in the northern part of the Gulf of Gabes on the Libyan-Tunisian border, is rich in oil and gas containing an estimated 3.7 billion barrels of oil and nearly 12 trillion cubic feet Tcf of natural gas. A Libyan-Tunisian Joint Oil Company (JOC), a 50-50 venture with the NOC and Tunisia's ETAP, has been set up to exploit the region. The Omar structure, on the Libyan side of the zone is estimated to contain more than 65% of the zone's total oil and gas reserves. The JOC awarded the entire block to a consortium consisting of Saudi Arabia's Nimr Petroleum (55%) and Malaysia's Petronas (45%) on February 1, 1997.

Sanctions caused delays in a number of field development and EOR projects, and to a certain extent deterred foreign capital investment. The full lifting of sanctions means that Libya now can resume purchases of oil industry equipment that would boost the national oil output.

Natural Gas

Libya's natural gas reserves are estimated at 46.4 Tcf, however these are mostly under-utilized and unexplored. Libyan experts place the potential reserves between 70-100 Tcf. The main gas fields include Attahadi, Defa-Waha, Hatiba, Zilten, Sahl, and Assumud. Libya wishes to increase the gas production both for the purpose of internal use, in particular power for generation, and for export to Europe. In order to expand the gas production, marketing and distribution, foreign participation and investment is sought by Libya. New sizeable discoveries have been made in the Ghadames, El-Bouri fields and in the Sirte Basin.

Ras Lanuf is one of the many dynamic points in Libya where companies from many different nationalities are working together

Libya also produces some liquefied petroleum gas (LPG), for consumption by domestic refineries. Current natural gas development projects include As-Sarah and Nahoora, Faregh, Wafa, offshore block NC-41, Abu-Attifel, Intisar, and block NC-98. A joint venture, a US$5.6 billion plan aimed at developing and exporting large volumes of natural gas to Italy, between Eni and NOC on the Western Libyan Gas Project (WLGP), is currently underway.

The WLGP plans to export 8 billion cubic meters (280 Bcf) of natural gas per year from a processing facility at Melitah to Italy and France over 24 years, commencing in 2006. This is to be done through a 370-mile underwater pipeline under the Mediterranean to south-eastern Sicily and the Italian mainland. Italy and France have committed to taking 140 Bcf and 70 Bcf respectively of Libyan gas. Libya also exports some liquefied natural gas (LNG).

Chemical Industry

An important economic strategy for Libya has been the effort to build on the achievements of its oil industry to create a well-developed chemical industry. Significant progress has been registered even if it has not developed as fast as originally intended.

Marsa El-Brega is Libya's main centre for the production of ethanol, ammonia and urea. The Marsa El-Brega complex is operated by the Sirte Oil Co. This site is owned by the National Petrochemical Company (Napectco). There is a methanol production facility at Al-Burayqah, and a petrochemical complex at Abu-Kammash. Another project is being developed at Ras Lanouf by the Ras Lanuf Oil and Gas Processing Company (Rasco) which should produce benzene, butadiene, methyl-tertiary-butyl ether and butane. There are also plans to produce LDPE, HDPE, PP and ethylene glycol at Ras Lanuf, as soon as finances permit.

Power Generation and Supply

Libya's power generation and distribution sector requires substantial investment and alternatives to public financing are being explored. The General Electric Company is responsible for the electricity supply industry in Libya. There is a separate corporation serving the city of Benghazi.

Plans and Developments

The Export-Import Bank of South Korea has agreed to lend US$ 99 million of the US$ 299 million required to expand and upgrade the 450 MW Benghazi North power plant. The project will convert the plant to a combined cycle system and double its capacity. Power production capacity is currently 4.6 GW. Most stations are oil-fired, though some have been converted to gas as have a number of thermal power stations on the coast.

Four power stations are being built and a 600km 225 kV interconnection with the Tunisian system of Societe Tunisienne de l'Electricite et du Gas (STEG).

There are also plans to develop other gas-fired facilities; these include a 450 megawatt (MW) gas-fired power plant in Sebha, an 800-MW power plant in Zuwara and a 1,400-MW power plant to be located on the coast between Benghazi and Tripoli.

A combined power and desalination plant in Sirte is also being planned. Power demand is growing at the rate of 6% annually and Libya plans to more than double installed capacity by 2010 at a cost of over $3.5 billion. Libya has expressed interest in developing nuclear power and Russia is willing to assist in this development.

Other national power projects include:

- 800-MW power plant in Zuwara on the west coast.
- 1,400-MW power plant on the coast between Tripoli and

Benghazi.
- 1,200-MW "Gulf Steam" combined power and desalination plant in Sirte.
- 650-MW Western Tripoli power plant.

Expansion and upgrading project at the 450-MW Benghazi North power plant doubling the plant's capacity and convert it to combined cycle.

- Expansion of the Az Zawiya power plant, west of Tripoli, by 300 MW.

MAIN COMPANIES

Arabian Gulf Company (AGOCO)

Giuma Mohamed Shaeb
Chairman
El Fallah, P.O. Box 693-325, Tripoli
Tel: + 218 - 21 - 4803880 / 5
Fax: + 218 - 21 - 4809881
E-mail: Info@agocoil.com
Website: www.agocoil.com

The Arabian Gulf Oil Company, established by law no. (115/71) issued by the revolutionary council on 7 December 1971, is an exploration company in which the shares of the British Petroleum Company (BP) were nationalized in concession no.65 covering the Sarir field. On 7 April 1971 the

The Al Zawiya Power plant, in the Western part of Libya

company started by operating one oil field (in Sarir) and one oil terminal (in Marsa El Hrega / Tobruk). Over the years, AGOCO has been extensively developed and now it undertakes the production and operation of eight oil fields, one terminal for exporting crude oil, and two refineries. The oil fields operated by AGOCO include the Sarir field, considered as one of the largest fields in the Jamahiriya and one of the ten largest fields in the world; the Messla field, considered one of the largest fields in Sirte; the Nafoora field, El Beda field and the Hamada field, which includes eight gathering stations connecting more than one hundred wells. AGOCO also operate the Sarir refinery, designed to refine 10,000 barrels of oil per day and the Tobruk refinery, designed to refine 20,000 barrels of oil per day and to cover part of the eastern part of the country's fuel requirements. Also, the Marsa El Hariega terminal, situated on the southern coast of Tobruk's commercial port, was inaugurated with the export of the first load of crude oil from the Sarir field on 10/01/1967. It has become quite apparent from the achievements of AGOCO over the years that the company has been extensively developed and improved from the outset of the initial company in 1971 due to its constructive foresight, keen workforce and informed management. The Arabian Gulf Oil Company has not only kept pace with the oil industry of today, it has become an accolade for the great Jamahiriya.

Azawia Oil Refining Company (ARC)

Eng. Al Muammari A .Swedan
Chairman
15715 Azzawiya - 6451, Tripoli
Tel: + 218 - 21 - 3610539 / 42
Fax: + 218 - 21 - 3610538
E-mail: infoazzawiya@azzawiyaoil.Com
Website: www.azzawiyaoil.com

Azawia Oil Refining Company (ARC) is a wholly owned subsidiary of the Nation Oil Corporation (NOC) and was incorporated under Libyan commercial law in 1976. The first refinery started production in 1974 with a capacity of 60,000 bpsd, producing LPG, naphtha, gasoline, kerosene (jet & d), gas oil, heavy fuel oil. In 1977 production capacity was doubled to 120,000 bpsd by operating the second refining unit. The refinery was designed to produce products to meet the latest specifications of international standards. It was prepared with environment preservation instruments to prevent air and water pollution. It also has all the safety requirements and is supplied with all associated units for utility needs like power generation, steam production, desalination units and air compressors as well as a tank farm for storing crude oil, both intermediate and finished products. The refinery has an oil terminal containing three offshore berths as follows: s.b.m (1): single mooring buoy first operated on 15/7/1974, used to receive ships of a capacity of 10,000 to 100,000 tonnes, s.b.m (2) : conventional berth first operated on 3/9/1974, used for receiving ships of a capacity of 5,000 to 30,000 tonnes of light products and oils lubricants, and s.b.m (3) single buoy first operated on 9/3/1977, used for receiving ships of a capacity of 15,000 to 140,000 tonnes of crude oil. The refinery contains two asphalt plants with a capacity of 100,000 tonnes/year each. These are as follows: the Azawia asphalt unit which started production in 1981 producing asphalt 70/60 and V.G.O and the Benghazi asphalt unit which started production in 1984 it producing asphalt AC-20, asphalt120-150, asphalt RC-250, asphalt MC-30 and V.G.O .The refinery also contains a lube oil blending unit with a capacity of 50,000 tonnes/year of the following groups of finished oils: gasoline engine, diesel engine, hydraulic oils and industrial oils.

Brega Oil Marketing Company (BOMC)

Naji Abdo Salam Bashir, 402 Tripoli
Tel: + 218 - 21 - 3600906 / 4808081
E-mail: Info @ brega-ly.com
Website: www.brega-ly.com

BOMC was established according to the law number 74 in 1971 after three companies joined together to make one company owned completely by the National Oil Corporation. These companies were Al-Watanya, Al-Brega and Al-Sedra. The company is established to deal with marketing and distributing petroleum products in addition of related commodities all over the country. The company created facilities to complete its activities which are: established storage tanks, gas stations and taking care of the operation and maintenance of these facilities, renting and operating petroleum tankers and ships in addition of the construction of pipeline networks for transporting the petroleum products. BOMC has storage trucks and rents others as needed for transporting operations of the petroleum products. BOMC made great efforts to develop and renew the storage facilities and increase the storage capacities for all the products to go with the development and consumption of these products, which go side by side with the development observed in the Jamahiriya in all fields of life. The most important products distributed by the company are lpg, benzene 95, domestic kerosene, aviation kerosene, diesel, heavy fuel oil, asphalt, synthetic oil and lubricants, gas cylinders, battery and radiator water. The company has put in effect a plan to face the local market needs from all kind of products and their derivatives. Therefore, it has established a number of modern storage facilities with a large capacity in different areas to provide marine facilities to be connected with these terminals. Also, it has established network terminal facilities in the airports to serve the planes. In addition, the company has established gas stations and domestic distribution gas centres all over the country to provide for the people's needs. The company owns a fleet of transporting tracks for fuel, lubricating oil plus gas cylinders. The company has established a unit of production of battery & radiator water for local consumption.

Jowef Oil Technology

15 Km Algeria Street, Benghazi
Tel: + 218 - 61 - 89505 / 89520
E-mail: Jowef @Yahoo.Co.Uk
Website: www.Jowfe.istop.com

Jowef Oil Technology is the only national company specialized in production and marketing of oilfield chemicals and providing equipment for oil and gas well drilling operations. Jowef also offers technical services in onshore and offshore oil and gas exploration and production. Jowef provides technical services for production companies in drilling and production wells in onshore and offshore by national qualified engineers including preparing proposals for drilling wells according to geologist reports for each well onshore and offshore and the collaboration with the client in all aspects of drilling operations. This includes providing wire line and production service for exploration and production wells by using modern electronic equipment in order to estimate the production power for exploration wells and determine problems in the wells, providing equipment service for mud cleaner equipment, centrifugal hole shaker, fitting tools for casing, tubing and spare parts for this equipment and also provide technical inspection of pipes, equipment and tools for drilling and production. Jowef also provides geological service and monitoring of the drilling operations in the exploration and development of wells in order to determine the type and the depth of the drilling beds which contain hydrocarbons and gas. Jowef, in doing so, owns and operates a number of plant producing chemicals which are required for oil and gas production and processing. The plant includes a grinding particulates plant (G.P.P): the plant produces barite bentonite calcium carbonates, and cathodic protection backfill materials. A majority of these materials are used in the oil and gas well drilling. The plant also includes a liquid products plant (L.P.P): this plant produces liquid specialty chemicals which are used in well drilling operations, crude oil treatment process. There is also a loss circulation materials plant (L.C.M): this plant produces lost circulating materials such as Micd wood fibre which are added to the drilling fluid to prevent the loss of drilling mud during the drilling operation. The Specialty Chemicals Department provides service in gas and oil treatment processing by using specialty chemicals such as: demulsifiers, corrosion inhibitors, scale inhibitors besides oxygen scavengers and treatment of cooling water systems.

Eni Gas B.V. Libyan Branch

Mr. Fuad M. Krekshi
Chairman
Dat El Imad Tower 1, Floor 16, Tripoli
Tel: + 218 - 21 - 3350746 / 8
Fax: + 218 - 21 - 3335672
Email: fkrekshi@enigasly.com
Website: www.eni.it

In 1974, Agip Company of Italy and the Libyan National Oil Corporation (NOC) reached agreements on 12 new discoveries. Those discoveries were made in the offshore field (N.C-41). The Boury field was the most important discovery and started production in 1988. With its aim to improve the economics of the project and to reach sufficient production to meet local Libyan market demands and to open new international markets for export, it decided to join the Wafa field in the Sahara desert and the offshore field, Bahr Essalam, to form a single project called "The Western Libya Gas Project" (WLGP). Due to the scattered locations of the WLGP's components in the sea, on the coast and the Sahara Desert and to its complexity, it has been subdivided in to eight main packages awarded to several major EPC/EPIC contractors for execution, in addition to the drilling activities (onshore and offshore). These main packages are the Wafa plants (desert & coastal), Onshore pipelines, Sabratha platform, Bahr Essalam Sub-sea systems, Bahr Essalam gathering and

export lines, Mellitah gas plant, telecommunication system and integrated automation system. The WLGP has an overall investment of 6.1 billion (3.1 billion for Eni) for the construction of treatment plants at Wafa and the city of Mellitah on the Libyan coast. This investment is also for the construction of onshore and offshore infrastructure, the laying of pipelines to transport both gas and condensates to the Mellitah plant and the laying of a 520 km-long, 32-inch underwater pipeline, the Greenstream gas pipeline, connecting Mellitah to Gela, in Sicily. The construction of the Mellitah plant was both a technical and schedule-driven challenge project. It was achieved two weeks ahead of time and has broken two world records: the deepest 32" pipe in the Mediterranean Sea and the biggest platform jacket lifted at one single time, 12000 tonnes on a single structure shipped from Korea by Hyundai. The Mellitah plant will have a treatment capacity of 10 billion cubic metres of gas per year; it has used more than 15500 workers from 8 different countries. Two billion cubic metres will be destined to the local market with the remaining 8 billion cubic metres being exported to Italy, through the Greenstream pipeline, to supply Edison Gas (4 billion cubic metres), Gaz de France (2 billion cubic metres) and Energia Gas (2 billion cubic metres). All the contracts are for 24 years and have "take or pay" clauses. The start up of these fields will make it possible to increase Eni's share of daily hydrocarbon production from 96,000 boe/day in 2004 to 250,000 boe/day in 2007.

This is a good example of close cooperation between a local and a Western Company, it should be proof of the confidence in safe investment in Libya

National Oil Fields & Terminals Catering Co.

Abolgasem A . Ben Shareia
Chairman
Airport Road, Tripoli
Tel: + 218 - 21 - 4801590

Fax: + 218 - 21 - 4802329

E-mail: Nat_cat@ hotmail .com

The incorporation of this company was deemed to be an important turning point for the oil sector, together with the necessity this activity constituted. After being a fertile ground for several individuals who practiced unfair working conditions two decades ago in order to realize the maximum possible profit, the moment came when the producers left in the September " Fateh" 1978 and subsequent arrangements led to the birth of the National Co. for catering pursuant to the resolution of the December 1981. In another turning point, activities were aimed at re-organizing this company in accordance with correct basis enabling it to realize its objectives the National Oil Company issued its resolution No. 45 for the year 1983 on re-organizing the company and forming its management committee. The national company for catering the oil fields and ports covers the largest part of oil sites due to its potential readiness and materialistic and human capabilities. These qualities enable it to fulfil its obligations towards the contracting companies and for them to surmount to the market effects, such as fluctuations and competition, and to be unshakable when facing all circumstances. The headquarters of the company is located in Airport Road, approximately 3 kilometres south of Tripoli, and it occupies within its premises an area of two acres. It consists of a three-floor administrative building plus stores for dry materials and other for fresh materials, the central refrigerator, the training centre, workshop of transport means maintenance in addition to a number of technical workshops and the site of safety and security, services, etc.

National Oil Wells Drilling & Workover Company

Mustafa Mohamed Idris

Chairman

P.O. Box 1106, Tripoli

Tel: + 218 - 21 - 3609830 / 33

Fax: + 218 - 21 - 4446743

The National Drilling Company was established in 1972 as a joint stock company. The National Oil Corporation (NOC), as a first step, owned 51% of the stocks, and Saipem Italian Company owned 49% of the shares, with a capital of LD 150,000. The mission of the company and its activities are to provide technical support, carry out all land and marine drilling services, including water wells drilling, and the maintenance of oil wells for all operator companies. Its mission is also the achievement of drilling prices balance and all other work associated with the drilling operations. In 1987 the General People's Committee issued decision n° 25 to join the National Drilling Company together with the National Company for Oil Wells Services, where the National Drilling Company was carrying out exploring activities and the National Company for Oil Wells Services was carrying out well maintenance in addition to other specialized works. The new company was called the National Oil Wells Drilling & Workover Company with a capital amounting to ld. (15,000,000) on 7/01/1987. In order to achieve these objectives, the company has prepared a qualified staff and owns a technical fleet consisting of seventeen maintenance for drilling and maintenance of oil wells which enables it to drill wells up to twenty thousands feet deep. The company is carrying out now about 45%-65% of the drilling activities in the Great Jamahiriya.

Petrocolmet Services Co

Gradimir Dragic

General Manager

P.O. Box 6135, Tripoli

Tel: +218 – 21 – 4777037 / 4776350

Fax: +218 – 21 – 4773398

Email: petrolcomet@petrolcomet.co.yu

Petrolcomet Services Co is a company based in Belgrade, Serbia & Montenegro, with its registered Branch Office in Tripoli, Libya. Petrolcomet, as a specialized engineering

شركة الزويتينة للنفط
ZUEITINA OIL CO.

Zueitina Oil Company
P.O. Box: 2134. Tripoli, Libya.
Tel: +218 21 3338011-14, 4441431-35
Fax: +218 21 3339109
Tripoli - Libya

and technical services company is organized to provide a wide range of technical services (process, mechanical, electrical, and instrumentation among others) for all types of industrial plants primarily related to hydrocarbon processing and production. The technical services offered by the company are provided on turnkey project basis as well as on contract/service. Orders requiring engagement of highly skilled and experienced specialists and experts to perform specific job assignments such as commissioning, start–up, turnarounds, maintenance and job training of client's personnel on various industrial plants. Since the inception of the company in the early 1990s it has been striving to provide and offer the highest degree of quality job performance. The management believes that this unrelenting and professional approach has helped develop the company into a sophisticated, world-class provider of technical services. Petrolcomet has developed strict working procedures that ensure best quality of service on a global scale.

Petrolibya Oilfield Services

Dr. Hadi S. Belazi
Managing Director
Ain Zarah, P.O. Box 7484, Tripoli
Tel: + 218 - 91 - 2123696
Fax: + 218 - 21 - 4625324
E-mail: petro_libya@hotmail.com

Petrolibya is a general oilfield contractor run by highly qualified Libyan personnel with decades of experience in the various aspects of the Libyan petroleum industry. The main objective of Petrolibya is to provide all the necessary services required by the operating oil companies; such as site preparation, landscaping, sand removal and stabilization, construction and maintenance of flow lines and pipelines, operation and maintenance of desalination, water and sewage plants, installation and maintenance of power transmitting lines and other electrical facilities and general maintenance and service of oilfield facilities. It also deals

with drilling water wells, seismic shot holes, cathodic protection and foundations. In association with others, the company undertakes the construction, maintenance and operation of early production facilities of small oilfields, the construction and maintenance of oil and gas tanks as well as design, supply and installation of cathodic protection facilities.

Petrolibya is planning to enter the well service sector; such as well work-over, testing, cementing and mud-logging. Petrolibya also acts as an agent or representative of manufacturers or service providers, who intend to enter the lucrative Libyan petroleum market, which will be the fastest growing and most rewarding sector in the coming years.

Ras Lanuf Oil & Gas Processing Co.

B. Zwary
Chairman
Ras Lanuf
Tel: + 218 - 21 - 3605177 / 82
Fax: + 218 - 21 - 3605174 / 3615242
E-mail: Info@raslanuf.com
Website: www.raslanuf.com

RASCO is an incorporated national company existing under the laws of Great Socialist People's Libyan Arab Jamahiriya (G.S.P.L.J), wholly owned by the National Oil Corporation (NOC). The company is governed by the rules and regulations of its foundation stated in the decision of the Company General Assembly No.1 of the year 1987. The company was created by the decision of General People's Committee No.137 of the year 1982 and amended by decision No.523 of the year 1986. The company is in charge of processing and refining oil and its derivatives, production of petrochemicals, plastics, fibres and any industry. Therefore it has been decided that the expansion of the complex is to be made in phases. Phase I, completed and commissioned in the eighties, is made up of a refinery with a capacity of

10,000,000 mta, an ethylene plant with a capacity of 1,167,000 mta, a harbour and shipment system as well as utilities and off-sites. Phase II is made up of a polyethylene plant with a capacity of 160,000 mta, a polymer handling plant (storage /bagging/wrapping) with a capacity of 240,000 mta as well as utilities, off-sites and interconnecting facilities. Also, the tank farm for the implementation of storage facilities dedicated to the petrochemical plants includes four covered floating roof tanks, nitrogen blanketed, with a capacity of 5,000 m3 each, for the methyl-tert-butyl-ether (mtbe) and benzene, two covered floating roof tanks, nitrogen blanketed, with a capacity of 5,000 m3 each, for the methanol, four pressure spheres, with a capacity of 2,000 m3 each, for butene-1 and spent butene. It also includes cryogenic storage for butadiene (2 tanks, with a capacity of 10,000 m3 each), flare system composed of two flares (one high pressure and one low pressure with an emergency ground flare), discharging 615t/hr (max). The high-pressure flare is for the phase II plants and the low pressure flare is for the butadiene storage facilities. The project extension of the hereinafter existing facilities in order to meet the requirements of phase II projects either implemented or future ones are as follows: Sea Water Desalination Plant No. 5, with a design capacity of 250 m3/hr; two nitrogen plants: gaseous nitrogen (design capacity of 3,000 Nm3/hr); liquid nitrogen (design capacity of 300 Nm3/hr). Other projects are: the hydrogen production plant (design capacity 300 Nm3/hr); air dryer unit (design capacity 5,650 Nm3/hr). New interconnecting facilities for projects envisaged under phase II, and provision of tie-ins for future projects, have yet to be implemented. We hereinafter expose the future envisaged plants as well their related production capacity, to be implemented as part of phase II projects, and for which extension to have been catered for under items 2, 3 and 4 above. More projects are the polypropylene plant (production capacity 120,000t/yr), petrochemical plants, being: butadiene extraction plant (design capacity 59,000 t/yr), butene-1 production plant (design capacity 18,000 t/yr) and a mtbe production plant (design capacity 55,000 t/yr) in addition to a benzene extraction plant (design capacity 88,000 t/yr).

Repsol YPF - REMSA

Mr. Graciano Rodríguez
General Manager
Dat El-Imad Tower 3 floor 5, Tripoli
Tel: + 218 - 21 - 3350380
Fax: + 218 - 21 - 3350381
E-mail: grodriguez@ryremsa.com
Website: www.repsolypf.com

The current activities of Repsol YPF E&P, carried out through its subsidiary Repsol Exploración Murzuq S.A. (REMSA), began in late 1994 after the signature of a contract with the National Oil Corporation of Libya. This contract was an agreement between the two parties as well as with two other foreign partners for Repsol to explore, develop and produce oil in the NC115 block located in the Muzruq region of the Great Desert, 800 km. south of Tripoli. Repsol Oil Operations (ROO) was assigned as the operator-company of that block, where more than one billion American dollars have been invested so far. A pipeline of 723 km has been constructed to link the NC115 block with the refinery located in the city of Zawia, and it is currently producing an average of 180,000 barrels a day with an infrastructure constructed for a total production capacity of 200,000 barrels a day. In November 1997 REMSA signed a second contract with the NOC regarding the exploration and production of two other blocks also located in Muzruq, where oil reserves were discovered and the NOC has already approved the development of that first discovery in the NC186 field. Since then, Repsol has managed to accomplish further agreements with the Libyan National Oil Corporation in order to expand jointly the exploration and production activities of Repsol in the country. Three important facts can give a quick overview of the importance and the fast development of Repsol in Libya. The NC115 block is one of the top quality producing points, Repsol is expanding its activities in the area of Muzruk and for the first time since the operations started, Repsol will explore outside Muzruk and offshore. Repsol bets on Libya as a country, the management pays special care and attention to the archaeological and environmental issues. Repsol sees real development possibilities in Libya and aims to grow in the oil and gas production.

Libya is a good business destination, but you need to make an effort and integrate the society and the culture, you need to give your best when coming to Libya.

Schlumberger Overseas S.A.

Mr. Fadel Fellah
General Manager
Dat El-Imad Tower 1, floor 7, P.O. Box 91931, Tripoli
Tel: + 218 - 21 - 3350060 / 61
Fax: + 218 - 21 - 3350064
Email: ffellah@tripoli.oilfield.slb.com
Website: www.slb.com

Since the two French brothers Marcel and Conrad created Schlumberger in 1927, Schlumberger Oilfield Services has become the largest and the leading supplier of exploration and production (E&P) services, solutions and technology to the international petroleum industry. With more than 45,000 employees in total and a presence in more than one hundred companies, Schlumberger Oilfield Services offers a suite of services and solutions that combine domain expertise, best practices, safe, environmentally sound wellsite operations and innovative technology. Its real-time technology services and solutions enable customers to translate acquired data into useful information and transform this information into knowledge as a basis for improved decision-making anytime, anywhere. Experienced in Libya since 1956, the company is present in eight different locations, having offices in Tripoli and Benghazi with

a total of 577 employees (345 Libyans and 75 foreigners). Excellently positioned in the Libyan market with a market share of approximately 70%, Schlumberger Overseas S.A. is technologically ahead and has a unique know-how of the country through its long experience. This enables it to be ahead of everyone in the Libyan market, especially when talking about the Great Libyan Desert where a thorough knowledge is essential to any success. Concerning the involvement of Schlumberger Overseas S.A., the company follows the group policy of diversifying its workforce, promoting from within and sending people overseas to implement and expand its personal experience and professional career. In 1999, Schlumberger Overseas S.A. launched its "S.E.E.D". Programme the Schlumberger Excellence in Educational Development, which is proof of the company's commitment to Libya. This program consists in connecting experts to youth, and youth to one another worldwide, bringing international standards of education into the country. For this purpose the first school was opened in Libya in September 2004, equipped with the latest technology to launch the program. With Europe as the biggest market for Libya, the country is a land of opportunities: now there is more access to reserves, 75% of the land is unexplored and the gas market is still underdeveloped.

Libya is definitely changing, the potential is high and the country is very well situated. Also, oil companies have to really believe in the human resources here in Libya. Schlumberger Overseas S.A. welcomes the newcomers and is ready to provide its services!

Saga Petroleum Mabruk AS (Hydro)

Tor Bjormulf Cund
General Manager
Dat El-Imad Tower 4, Level 10, Tripoli
Tel: + 218 - 21 - 3350344
Fax: + 218 - 21 - 3350343
E-mail: tor.b.lund@hydro.com

Website: www.hydro.com

Hydro is a Norwegian energy and aluminium supplier operating in more than 40 countries. It is the leader in offshore producing of oil and gas and the world's third-largest aluminium supplier. Listed on the New York Stock Exchange and with 36,000 employees worldwide, the company is also a leader in the development of renewable energy sources. In Libya, Hydro is a partner in the Mabruk field, in the Sirte basin, and in the NC 186, 187 and 190 concessions in the Murzuq basin. The company has also achieved an alliance with the German oil and gas producer Winthershall in nine areas located in the eastern part of Libya. Hydro has been present in the country since 1994 and it was registered as Saga Petroleum Mabruk AS in 1996, as a branch office of Hydro, but the company is expecting to change the name to Hydro in the next months to come. Hydro not only has its oil and gas experience in the country, the world's largest supplier of plant nutrients YARA had a long business relation with the country until march 2004 when the company divested from Hydro. Operations in Libya represent at the moment only 1% of the Hydro group operations as it is only present as a shareholder together with some of the most important international oil and gas companies. However, the company is aiming to become an operator and become one of the leading companies in the years to come. Libya has large potential resources and volumes of oil and gas, very easily brought to market due to its high quality. Libya is also a very interesting country from a commercial point of view as it has a strategic location within the Mediterranean Sea, and this very interesting for the gas industry.

If you want to do businesses in Libya don't expect a profit the following day, investment in this country requires a long time and a long effort, in order to succeed you must be patient enough

Sirte Oil Company

Ahmed Hadi Aoun
Chairman, Tripoli
Tel: + 218 - 21 - 3610376 / 90
Fax: + 218 - 21 - 3610705 / 3605118
E-mail: aaoun@sirteoil.Com
Website: www.sirteoil.com

Sirte Oil Company (SOC) is an operating company located 800km east of Tripoli in the town of Marsa El Brega. It is one of the largest operating subsidiaries of the National Oil Corporation of Libya. SOC's diverse operations include the production, manufacturing, transmission service and exportation of crude oil and natural gas. Its operations also include the extraction and processing of lpg and naphtha, the manufacturing of methanol, ammonia and urea and the liquefaction of natural gas (lng) for domestic and European export markets. The oil and gas production quantity varies in the different field locations which SOC produces from. Some of the fields are producing whilst others are underdeveloped. The fields operated by SOC are classified as follows: oil and associated gas fields (Nasser, Raguba, Lehib, Jebel, Arshad, Ralah, Wadi, and Ain Jarbi); non - associated gas fields (Hateiba, Sorrah, Mehgil, Sahl, Assamoud, Attahaddy); petrochemical plants; two methanol plants with a total capacity of 2000 mt/d; two ammonia plants with a total capacity of 2200 mt/d; two urea plants with a total capacity of 2750 mt/d; an lng plant of a capacity of 120mbpd of lng and 3mbpd of lpg and 40 mbpd of naphtha. The refinery has a capacity of 8,400 b/d and an initial start-up in 1963 producing gasoline, kerosene, jet fuel, naphtha, ado and heavy fuel oil. The major current projects are the Attahaddy Field Development Project, the El Khoms - Tripoli 34" gas pipeline, Sidra & Washkah booster stations and the Tripoli-Melita 34" gas pipeline. This last project involves processing 350 mmscfd of raw gas, which will yield approximately 270 mmsced of gas and 30 mbd of condensate. The El Khoms - Tripoli 34" gas pipeline and Sidra & Washkah booster stations involves the extension of the coastal gas pipeline from El Khoms to Tripoli in order to make natural gas available to the South Tripoli power station and in the future to the Souk Khamis cement plant and the West Tripoli power station. The Tripoli - Melita 34" gas pipeline involves the connection of the coastal gas pipeline network thus allowing more flexibility to the existing network by supplying 200 mmscfd of natural Eni gas Melita Facility (based in the west of Libya) to Zawia and Tripoli power stations. This will create 73 Job opportunities for technicians and operators in the region. Among the major executed projects we find the Brega - Zwetina 34" gas pipeline, the Zwetina - Benghazi 34" gas pipeline, the Brega - El Khoms 34" gas pipeline and the Attahaddy - 91.5 km gas km gas and condensate pipelines.

Veba Oil Company

Saad Naji Zaid
Chairman
Janat Al-Arief / Bashir Assdawi St. Tripoli
Tel: + 218 - 21 - 3330081
Website: www.vebalibya .com

On 11/03/1988 in Holland, a joint contract between the National Oil Corporation and Veba, the German company, to establish a company concerned with oil exploration and production was made, having a branch in Jamahiriya under the name Veba Oil Operations Company. The National Oil Corporation held 51% of the shares while Veba held the additional 49%. Complete supervision and operation of previously owned Mobil and Gelnsberg Company sites and fields were taken on with this agreement. Veba Oil Operations Company works through its site on the Libyan coast, owning the largest oil port working in the Jamahiriya, exporting about two thirds of the Jamahiriya production through Ras Lanuf oil port. This was in addition to exporting production of neighbouring fields and of five fields of its own. This distinguished position makes it one of the most important companies working in the field of exploring, production and exporting of crude oil. Since its establishment the company has made a re-evaluation and study of all its privileges in order to know the size of the strategic reserve of the producing fields. This included the setting out comprehensive studies to develop all the utilities, in addition to longitudinal expansion in seismic and the exploration and redevelopment of the producing wells and protection of the underground reserve from rapid depletion. This is done by adopting advanced techniques and highly modern technologies by injecting huge water quantities inside the Al-Ghani field to the depth of the oil reserves to maintain the size of the originally produced crude and increase in producing the largest possible quantity inside the reservoirs. Regarding the exploration side, the company has spent huge amounts on exploration programs within its approved privileges, enabling it to maintain its constant production size over the years. Its assignment to operate the Al-Naqa field in 2002 has caused a rising performance level in such a way that secures continuity in raising its production averages. Pumping water in Al-Ghani and Amal fields, and maintenance of pipelines 30-36 inches is one of the most important projects executed by this company at present. Also, it is involved in the adoption of complete maintenance programs for all its sites and oil fields together with promoting and developing the lines and pipes networks through hundreds of kilometres as well as the large expansion in constructing modern and supporting utilities, modernization of its airports and runways to be used by the company's aircrafts working between its fields. In addition, it complies with periodical safety programs and maintains environment protection together with execution of plans and programs leading to promotion of the human element; in providing full health care in all various sites of the company.

Waha Oil Company

Abdussalam milad amari
P.O. Box 395 (off airport road), Tripoli
Tel: + 218 - 21 - 337161 / 3331116
Fax: + 218 - 21 - 3337169 / 3330985
Website: www.wahaoil.com

Waha Oil Company is one of the major national companies engaged in the field of oil and gas production. Its incorporation date was in 1955, when the Oasis Oil Company was incorporated as an operator for three American companies. Waha Oil Company started its operations on Libyan soil in 1956 and succeeded in discovering oil in commercial quantities in the area north of Bahi oil field. The company continued its successful operations after that date, and in 1960, the Waha and Defa oil fields were discovered and later in 1961 the Gialo and Samah oil fields were found. In 1962, the company was successful in discovering the Zaggut, Belhedan, Khalifa, Balat, Masrab, Harash and Al-Faregh oil fields. On 22 may 1962, the company's first oil shipment of crude oil from its fields was exported to the world oil markets via the company's Es-Sider oil terminal. In 1969, in the famous oil liberation revolution, the state of Libya gained full domination over the oil sector, and in 1973, a joint - venture agreement was signed between the National Oil Corporation (NOC) and the Oasis Group of companies, under which the NOC held 51% of the company's fixed and movable assets, and for the first time, a board of directors was appointed for managing the company's affairs. The board was made consisting of three members of which two represented Libyan interests and one represented foreign companies' interests. The fourth member in the Waha Oasis Group, the Dutch company Shell, with a different stand on the process, abstained from signing the agreement. Consequently, its share in the group was nationalized in 1974, resulting in the loss of its full rights under which the NOC owned 59% of the company shares. In 1980, the American administration imposed unjust sanctions on the Libyan people with the oil sector as the main objective. The American government ordered its oil companies operating in the great Jamahiriya to leave Libya, as well as prevent its citizens from working in Libya, and furthermore, imposed heavy fines on those violating such orders. Following such acts, a decision was issued by the General People's Committee under no. 350 establishing the Waha Oil Company as wholly owned by the NOC. Since that date, the Waha Oil Company continued its activities in order to increase the oil reserve, enhance added oil recovery and train and develop national staff in the company's different areas of activities. The company operates a number of major oil and gas producing fields and exerts strenuous efforts in the development of existing fields and exploration of new ones. The company's major oil fields are as follows: (1) Waha, (2) Gialo, (3) Dahra, (4) Samah, (5) Al-Faregh, in addition to the Es-Sider oil terminal. The Waha Oil Company has always been the first to avail of modern technology used in the oil industry, such as control of oil fields and Es-Sider terminal operations, also used for the control of power transmission lines and for material movement in all company warehouses. The company also owns and operates an advanced communications system connecting all company locations, together with the general post network. Local and international standards are applied by the company. Future Projects include: seismic surveys of approximately 6,000 sq.km, drilling of 25 exploration wells in the areas allocated to the company and continuation of drilling of designation wells for the recent discovery at Samah in addition to several other activities aimed at enhancement of performance of the company's operations.

Wintershall Libya

Dr. Ties Tiessen
General Manager
Corinthia Bab Africa, Commercial Centre, Tripoli GSPLAJ
Tel: + 218 - 21 - 3350135
Fax: + 218 - 21 - 3350136

MAIN SECTORS OF THE ECONOMY: ENERGY

Email: ties.tiessen@wintershall.com

Website: www.wintershall.com

Wintershall started its operations in 1958 and never left the country even during the embargo. Wintershall specialises in energy and is a wholly owned subsidiary of BASF AG of Germany. Together with its affiliates, Wintershall is active in exploration, production and trading of crude oil and natural gas. In addition, the Kassel-based company markets capacities for crude oil and natural gas storage, natural gas transportation and optical fibres. Wintershall has been active in the exploration and production of oil and gas for over 70 years Europe, North Africa and South America, as well as Russia and the Caspian area. The company employs more than 1550 staff from more than 30 countries. In Libya, the company operates five fields and is, together with the National Oil Corporation and Total, a joint owner of in an offshore concession. Also, Wintershall has designed and constructed a gas utilisation plant capable of recovering from the produced oil, an estimated one billion cubic meters per year of lean natural gas and 275,000 tons of gas condensates. Wintershall is also committed with the country to several different aspects apart from its economic activity. The company has highly invested on environmental projects such as a water plant and is very aware of its social obligation to Libya, initiating the Libyan integration and development program, designed to increase the number of skilled workers within the company both in Libya and abroad. Regarding the arrival of new oil companies to the country, Wintershall claims to be maintaining its position in the top oil companies in the future as it has a long experience and a good geological knowledge of Libya, this added to its quality and technological advantages. When describing its Libyan experience, the company claims that it has always been happy to be collaborating with the NOC as they have a very professional and well-educated staff. The EPSAs launched by them meet the EPSA international standards and Wintershall will continue participating in them in the future.

The Great Libyan Desert covers a land reach in oil and gas

In order to succeed in Libya, always say what you think and do what you say

evident everywhere you look, which further demonstrates the resilience of the Libyan people. We are confident that foreign investors will use this opportunity to work together with us to further develop our country to its full potential

Zueitina Oil Company

Mohamed M. Own
Chairman of the Management Committee
P.O. Box 2131, Tripoli
Tel: +218 21 3338011-14/4441431-35
Fax: +218 21 3339109
Email: moun@zueitina-ly.com

Zueitina Oil Company was incorporated in 1986 as a Libyan owned Company with a mandate to carry out the whole range of oil operations, based on the frozen assets of Occidental International, which was established in 1966. Zueitina Oil Company currently operates on behalf of the National Oil Corporation of Libya (NOC), Occidental International and OMV of Austria, which acquired a 25% share of Occidental International's interest in 1985. In the year 2003, NOC was authorized to negotiate standstill agreements with American companies. NOC is now in the final stages of discussions for the return of Occidental International. The mandate of Zueitina Oil Company is to efficiently produce Crude Oil, Natural Gas and Condensates at a minimum cost, with an optimum recovery, and with a great emphasis being placed on environmental and reservoir protection. The Company is currently handling up to 20% of the country's Oil and Condensate exports. Zueitina Oil Company also enjoys the competitive advantages of having a fully developed infrastructure, from Drilling to Terminal Facilities, and employs a dedicated team of experienced and highly qualified engineers and technicians.

We consider our country to be blessed with many unique features, not the least of which are the people, who are competitive, but easy going. Libya is a very safe place to live, and is also a very safe place to invest in. Now that the sanctions against Libya have been lifted, and the embargo is behind us, investor confidence has quickly returned and is

Local **solutions**
global **expertise**

FINANCE

MAIN SECTORS OF THE ECONOMY: FINANCE

OVERVIEW

The Structure of the Banking Sector

The banking sector in G.S.P.L.A.J. consists of: the Central Bank of Libya and nine commercial banks. Three of these commercial banks are government controlled banks, including The National Commercial Bank, Al Algumhuria Bank and Umma Bank Represented by the Central Bank of Libya. Two of the other commercial banks are Alwahda Bank and Sahari Bank, more than 80% of which are owned by the government. The last four commercial banks fully owned by the private sector and are the Trade & Development Bank, Aman Bank for Trade and Investment, Arab Universal Bank and Alwafa Bank. These commercial banks have more than 330 branches and banking agencies spread all over the Great Jamahiriya as well as 48 national banks fully owned by the private sector which pursue their activity within the geographic area in which they are located. There are also three specialized banks including the Agriculture bank, the Real-estate Savings & Investment Bank and the Development Bank. As for the Libyan foreign bank which is fully owned by the central bank of Libya, is pursuing its activities outside of the Great Jamahiriya.

The Significance of the Banking Sector

The banking sector in Libya plays a vital role in the national economy through its services as well as its financing of the different economic activities. It provides credit and loans necessary for companies, public establishments and national sectors. It also finances a great deal of supplies of consumer goods and the capital goods required for economic promotion and development. It grants the real-estate loans and social loans for society's individuals in order to raise their living standards. By the end of 2003, the assets of the Libyan banking sector amounted to 16,409 million dinars, at an annual growth rate of 4.1% in 2003, while their deposited liabilities amounted at about 10,204.0 m/d at an annual growth rate of 7.5%.

The Realized Achievements

Monetary Policy

In its efforts aimed at promoting economic stability and levels of national economic performance and preparation of the appropriate circumstances for foreign investment flow, the Libyan Central Bank took the necessary steps, coordinating with the competent authorities, to adjust and Unify Libyan dinar exchange prices against foreign currencies devaluing its official dinar exchange rate by 50% in 2002. This policy has given fruit to positive results and indicators with regard to national economic performance, which was praised by many international financial institutions. As a result of the real gross national product released in 2003 a growth rate of 3.7% was obtained and thus there was a rapid increase in the Libyan dinar value against neighbouring countries' currencies. Currency supply witnessed a growth of 9.3%. The national economy was able to absorb this growth through the larger growth of the goods and services in supply, which led to a rapid decrease in the annual standard rate number for prices (1999 prices) for the fourth year in a row.

Banks Reorganization and Privatization

With an aim at trust and transparency support in the banking sector, a committee has been formed to study and reorganize commercial and specialized banks activity. This committee has completed its task and a relevant report was prepared which included a number of recommendations and suggestions in the field of bank status remedy with regard to capital, ownership, management banking supervision, number of banks and banking services. The Central Bank of Libya, for its part, formed two technical committees charged with the task of supervising the privatization of two commercial banks to the national sector. The two banks are Sahari Bank and Wahda Bank.

Banking Services

Prompt efforts are being exerted to promote and modernize the banking sector in the Great Jamahiriya and to diversify its scope of services in quantity and type as required by economic activity and in pace with international developments and events. In this regard, the work is now being done to create a national system for payments, which is expected to play a vital role in promoting and diversifying the banking service and especially in the utilization of new payment means such as using the credit card, cheques clearance operations and money transfers. The Central Bank of Libya gave permission to commercial banks to issue credit cards in foreign currency for individuals (paid in advance) at a maximum US$10,000 annually. With regard to banking sector staff qualifications, there are training programs, organized and implemented locally and abroad to promote staff knowledge. In addition, the banking training centre is being promoted and changing into an institute for banking studies in cooperation with some specialized international banking institutions.

Commercial and National Banks Control

The Central Bank of Libya carries out its role of controlling and supervising commercial and national banks in accordance with the powers invested in it by virtue of the Banks, Money and Credit Act no. 1 for the year 1993, with the goal of achieving stability of economic activity, ensuring the safety of banks' status and depositors. The banking control and supervision takes different forms and norms, some of which were being performed in the office where reviewing

Another view of the Dat El-Imad complex, home of many banks of Libya.

and analyzing data supplied to the Central Bank, such as financial centres data and monetary and banking statistics of the commercial and national banks is performed. Other control and supervision actions are carried out through periodic field inspections, unexpected inspections, and the specific inspections of specific cases of banking dealings. The bank also closely monitors the safety of financial transactions and dealings of commercial banks with regard to its commitment to implement international standards.

Loan Granting

The commercial banks play an essential role in economic and social program development financing through attracting the different kinds of deposits and savings. They are involved in production and service investments, including all economic activities and goals. As part of its contribution to solving the problem of providing appropriate housing for citizens, the commercial banks, over the past three decades, have granted about 129,000 loans for housing valued at 2,200.3 million Libyan dinars, in addition to its contribution to the housing program which amounted to about 510.0 million Libyan dinars. The fund for outstanding real-estate loans at the end of 2003 is set at about 1,472.1 million Libyan dinars, or a ratio of 21.7% of the total credit granted by commercial banks. In order to join the international finance and banking developments and events, the Libyan banking sector commenced developing and diversifying its scope of services carried out. It realized the following procedures and resolutions:

- Permitted the commercial banks to issue foreign currency credit cards at the power of its clients deposit accounts valued in Libyan dinars or in foreign currency kept by individuals and organizations with this banks.
- Approved raising the capital of some commercial banks with the aim of raising the rate of the quality of capital of these banks.
- Established a core for a money market in the Great

Jamahiriya and initiating the sale and purchasing of bonds and shares.

- Prepared a draft law concerning reorganization of banks, money and credit to replace the law no. 1 for the year 1993, in order to remedy many issues. Bank dealings existing in banks are to done in a manner which takes into consideration the local economic and social development and international standards.
- Implementation of the required arrangements to prevent money laundering through the banking sector by creating a financial information unit in the Central Bank of Libya and similar units in the commercial banks. These units are to carry out observation and follow-up all dealings and transactions which are suspected of having to do with illegal dealings or money laundering. The Central Bank of Libya aims to fight money laundering and to transfer any such activity to the competent authorities.
- Published more economic and statistical data and information concerning the banking sector in the annual and the quarterly reports as well as on the Bank's website.

The Future Theory

Through the general framework of strategy adapted by the banking sector in Libya, the future banking prospects can be summarized in the following points:

- Development of the current financial institutions and creating newly developed financial and investment institutions which could play a vital role in the primary market with regard to marketing, distributing and managing of the new issues for the new joint stock companies.
- Driving toward expanding the ownership base in the general commercial banks and creating competence among such banks with the aim of promoting the quality of staff performance and releasing a proper financial gain.
- Modernizing the Libyan banking industry and driving toward projects assessment, providing financial and consultation services and playing the role of the financial broker

between the saver and investor, and creating investment funds which contribute to transferring individuals into shares for owning part of the capital of the companies offered for privatization.

- Promotion of banking legislations in a way which allows transfers from commercial banks, whose activity is restricted to limited duties, to the capacity of the comprehension bank, which is able to enter into and manage financial and new special investments toward the promotion of the role of future financial market.

Agreements for Investment Promotion and Guarantees

Agreements for Investment Promotion and Guarantees held between the Great Jamahiriya and a number of countries around the world

1	Italy	2	Belarus	3	Ukraine
4	Cyprus	5	Switzerland	6	Morocco
7	Austria	8	Portugal	9	Malaysia
10	N. Korea	11	Serbia and Montenegro	12	Germany
13	Belgium	14	Algeria	15	Malta
16	S. Africa	17	Chad	18	Egypt
19	Bulgaria	20	Iran	21	Jordan
22	Magreb Union	23	Syria	24	Croatia
25	Tunisia	26	The Sudan	27	Niger
28	The joint agreement for capital investment concluded within the framework of the Arab league.				
29	The agreement concerning settlement of investment dispute.				

Agreements for Investment Promotion and Guarantees still under discussion with some countries

1	Vietnam	2	Cuba	3	Macedonia
4	Holland	5	Pakistan	6	Yemen
7	Albania	8	Denmark	9	S. Korea
10	France	11	Indonesia	12	Bahrain
13	Kuwait	14	Romania	15	Check Republic
16	Poland	17	Slovakia	18	Congo
19	Turkey	20	Iraq	21	Russian Federation
22	Zimbabwe	23	Guinea Conakry		

Present Situation of the Foreign Investments:

Found here below is a statement of the existing foreign investments under law no. 5 of 1997 since its issue and until the middle of 2004.

* The investment costs of the projects have reached the sum of L.D. 1,375,161,653 of which the sum of US$ 918,859,032 (L.D. 1,194,516,746) is foreign capital and represents a ratio of 87% of the total investment built.

* The following table shows the details:

S.L No.	Field of Investment	Projects		Investment Value of the Projects				Foreign Share in the Invest. Field
		No.	%	US$	Equiv. To L.D.	Total in L.D. (*)	%	
1	Industry	46	60%	485,352,872	630,958,736	719,940,971	52%	88%
2	Tourism	5	6%	275,690,050	385,527,066	416,147,070	30%	86%
3	Agriculture	2	3%	50,500,000	65,650,000	65,650,000	5%	100%
4	Services	9	12%	50,508,395	65,660,914	76,262,914	6%	86%
5	Health	12	16%	47,939,252	62,321,028	84,496,796	6%	73%
6	Marine Wealth	2	3%	8,768,463	11,399,002	12,663,902	1%	90%
	Total	76	100%	918,759,032	1,194,516,746	1,375,161,653	100%	87

* The foreign investment evaluated in dinars plus the share of the national investment in L.D.

(The exchange rate is US$ 1 = 1.3 L.D.

* The following table shows the geographic distribution of investment projects capital resources as per the continents.

Continent	The Investment Value of the Projects		Ratio %	Remarks
	US$	Equiv. L.D.		
Europe	633,237,702	823,209012	68.9%	Number of the invest. Countries (13). Switzerland share is 43%
Africa	161,143,054	209,485,970	17.5%	Number of the investing countries is (4) countries. Egypt (56%), Tunisia (18%), Libya (20%), Morocco (6%)
Asia	67,978,276	88,371,758	7.4%	Numbers of the investing countries is (6) states. Saudi Arabia share (43.9%), UAE and Jordan (25.1%)
North America	50,000,000	65,000,000	5.5%	Number of the investing countries is (1) which is Canada
South America	6,500,000	8,450,000	0.7%	Number of the investing countries is one which is Brazil
Total	918,859,032	1,194,516,741	100%	

MAIN COMPANIES

Libyan Arab African Investment Company

Dr. Rajab M. Lasswad

Director of the Planning Department

P.O. Box 81370-76351, Tripoli

E-mail: lasswad@laaico.com

Website: www.laaico.com

From humble beginnings in the mid-seventies, LAAICO business interests now span the four corners of the African continent in more than 25 countries and in diverse sectors from hotels & real estate, industry, agriculture, trade to mining and telecommunications, never wavering in its convention in Africa's potential for growth and development. The goals of the company are to generate revenues for the shareholders, to promote inter-African trade, to promote the exchange of technology and to integrate the African countries economies, regionally and globally. In addition, the company's policies are to invest in economically viable projects, to diversify the investments, both sector-wise and geographically, to practice a sustainable utilization of the environment and to utilize the local raw materials in a profitable and professional manner. Confident of the endless potential for business growth in Africa and backed by past experience, LAAICO is planning to launch an ambitious investment program for the next five years aimed at the expansion in the fields of telecommunications, mining, manufacturing and tourism. In doing so, LAAICO intends to enter into strategic alliances with capable partners attracting both foreign capital and know-how to Africa. Currently, LAAICO is present in Gabon, Congo, Burkina Faso, Niger, Mali, Guinea, Chad, Uganda, Benin, Liberia, Ethiopia, Nigeria, Madagascar, Central Africa Republic, South Africa, Eritrea, Zimbabwe, Zambia, Rwanda, The Gambia, Ghana, Togo, Comoros and the Democratic Republic of Congo.

Libyan Arab Domestic Investment Company

Mr. Salah Bilasher
Chief Executive Officer
Tripoli
Tel: + 218 - 21 - 4807042
Fax: + 218 - 21 - 4807021

The company was created in 1993 with an initial capital of one million Libyan dinars. This capital figure was later increased to ten million Libyan dinars. With up to 84,000 private shareholders, LADICO has shares in many other companies, but most importantly, it completely owns six investment companies located in Tripoli, Misurata, Sirte, Gherian, Benghazi and Botnan. LADICO has a total staff of more than 1200 employees. The company is present in three different fields of activity: the car industry, the construction industry and more recently in the tourism industry. Regarding the car industry, the first activity of LADICO was to sell up to 100,000 imported Asian cars among others. It also set up a network of maintenance workshops and spare parts stores. In addition, the company has shares in the Jamahiria Cars Company Sajico, which, together with Daewoo is to build the "libo", the first Libyan car. However, construction is currently the most important activity of the company, which has shares in 10,000 housing units of the 60,000 that the government is

building. LADICO is also taking part into the construction of 800 units. The company has its own equipment for the construction sector and has also been involved with many important projects such as building hospitals and desalination plans. In the tourism field, the company just finished building and started operating a brand new tourist resort in Misurata. The resort is located on 54 hectares of land owned by the company of which only 12 hectares are developed, which means that there are another 42 hectares to be developed in the future. Each house has an average of 80 square meters and 190 of them have double bedrooms. In the future the management hopes to build more tourist villages like the one in Misurata and get more housing contracts as well as assembling cars in addition to importing cars to Libya.

"We are ready and prepared to work in any kind of housing project. The same is true with tourism and we are willing to accomplish future joint ventures with foreign investors coming to Libya"

Libyan Foreign Investment Board

Rajab M. Shiglabu
Chairman
Ben Ghashir Road, 20
Tel: + 218 - 21 - 3618686 / 3609613
Fax: + 218 - 21 - 3617918
E-mail: rajab@investmentinlibya.com
Website: www.investmentinlibya.com

In accordance to article 5 of the law 5 enacted in1997, the Libyan Foreign Investment Board was set up with the main objective of implementing the law itself. The board was set and started functions between 1999 and 1998. The first start for investment projects approvals was in the year 2000 with four projects, two of them were the industrial Tajura car factory project and the Bab El Africa (Corinthia) tourist project.

The way the board works is the following: first, the foreign investor comes with an idea, then a preliminary application with the details of the nature of the project is done. Then, after an initial approval seeking to check that the project meets the law, a final application is required for a final approval. Mainly, the final application will require information on the individuals or on the organisation in the case it is a company that wishes to realise an investment project. Once all the process is done, the board will assist the foreign investor in order to submit the last documents and get the final approval by the office of Secretary of the General People's Committee for Economy and Trade, which will decide and issue the necessary permits. The board will also assist the investors by facilitating the help regarding the working visas. It will also follow up the setting up of the projects, helping with the relations with the local authorities. It will enable the new investors to get the incentives offered by law 5. These incentives are an exemption from custom duty on the equipment and raw materials the company wishes to bring inside the country for the starting operations as well as a five years income tax holiday among other incentives. Whenever an investor might be interested in a country, he will look at three factors, resources to set up a project, size of the market and productivity or efficiency to produce at the cheaper cost. For the first factor, Libya should be interesting for energy and tourism resources as well as raw materials for construction, PVC and metals among others. Concerning the second factor, Libya offers ties with the Arabic world as well as the African Union and is very close to southern European countries. It is important to consider that Libya offers security, there are no religious conflicts inside the country and Libyans are used to foreigners. For further information please see the Board website.

"Libya gives a lot of opportunities and chances for those interested in the field of business, opportunities you shouldn't be let go and we are here to help anybody!"

Libyan Insurance Company

Dr. Salem M. Bengharbia
Chairman of the Board
Tripoli
Tel: + 218 - 21 - 4444179
Fax: + 218 - 21 - 4444178

Libyan Insurance Company was founded in 1964. With an average premium per year of 150.000.0000 Libyan dinars, the company provides all types of direct insurance. Among the variety of the services offered are car insurance, fire and marine accidents insurance, travel insurance and health insurance among many others. In addition, the company deals with the main reinsurance companies in the world. Currently, the company is publicly owned but the management is working on the possibility of offering an indeterminate amount of shares to the Libyan private sector. The company is also involved in current tourism and investment projects in the country. The Libyan Insurance Company is one of the biggest insurance companies in the northern African region as well as in the Arab world. With a capital of 50 million Libyan dinars and it is the first insurance company established in the country. It is definitely the most important insurance company in Libya as well as the most experienced one. Currently, the company is working really hard on improving its services as well as improving customer care and developing new agreements locally and worldwide.

"If you come to Libya you will get good results. I invite tourists and visitors just to come to visit the country; there are many opportunities in very different fields. Business here will be very successful. Libya has very nice places and very nice weather. We have everything that anyone is looking for."

Main Sectors of the Economy: Finance

Sahara Bank

Dr. Salem M. El Ghumati
Chairman
P.O. Box 270 Tripoli
Tel: + 218 - 21 - 330724 / 3339265
Fax: + 218 - 21 - 3337922
Email: sahbankit1@lttnet.net

Before 1964, the Sahara Bank was formerly known as Sicilia Bank. It was a union of three shareholders: Libya, Sicilia Bank and Bank of America. Currently, the bank is 82% owned by the Central Bank of Libya and 18% owned by private share-holders. It is with Wahda Bank, one of the two banks owned by the Central Bank of Libya, that it is going under a privati-sation process. The Sahara Bank is ranked as the first Libyan bank and ranked as one of the top one thousand banks worldwide; this is due among other facts, to its strong equity capital. With close to 16,000 employees, the bank is 44 branches strong in Libya and has correspondent banks in the major European cities. The competitive advantages of the bank are the quality of the services it provides to its cus-tomers and the interesting financial settlements offered for loans and credits. In a social perspective, the Sahara Bank trains its staff locally and overseas and has contributed to Libyan society with loans and credits granted with favourable interest rates. Due to its structure and position in the Libyan and international bank rankings, Sahara Bank's major asset towards the foreign investment coming to the country is the confidence it may offer to the foreign investors and foreign companies that are coming to Libya. The Sahara Bank has not only experience with big local companies but also with foreign ones, such as some of the oil and gas foreign com-panies operating in Libya among others. The confidence and quality of services offered by Sahara bank, together with the safe climate, big business opportunities and good geo-graphic location that the country enjoys should be essential to attract foreign capital in the years to come.

"When coming to do business here, choose Sahara Bank to be your partner!"

Umma Bank

Ayad S. Dahaim
Chairman
P.O. Box 685, Tripoli
Tel: + 218 - 21 - 3331195 / 3332888
Fax: + 218 - 21 - 3330880
Website: www.ummabank.com

In 1907 the Banca di Roma was set up in Tripoli in order to pave the way for the colonisation of Libya by Italy. Then, on the 13th of November 1969, the Libyan Revolution Council gave 50% control of foreign banks to the Libyan Central Bank. The following year, the Banca di Roma became Umma Bank ("the Bank of the Nation") and the Central Bank became its sole owner. In the beginning, the Umma Bank was very small and concentrated in the northern coast of the country. Now, it has 54 branches all over the country and cor-respondence with most of the main banks overseas. The Umma Bank is currently dealing with SWIFT codes and deals with approximately 60% of foreign trade. Among the clients of Umma Bank are many international companies that rely on the good quality of the services provided and the speed of the cash transfers done by the bank. The time factor is essential for any commercial business as "time is gold". The bank is already working on the introduction of electronic credit cards. The management would like to not only be a facilitator of the VISA or AMEX services through ATM machines, but also be a card issuer in the near future. Among the future projects of the bank is a project to open a branch in the same location as the Corinthia Bab Africa International Hotel in order to facili-tate the access of their services to the current and potential clients. The Umma Bank is not under a privatisation process at the moment as the Central Bank is only privatising two of their five banks for the time being, but the privatisation process could also happen soon for the Umma Bank. In the future, the Umma Bank would like to expand its correspondent network and grow in and outside Libya, especially within the African Union. It already has links with some of its neighbouring countries. In the future, the management would like to see the Umma Bank as a leading bank in Libya.

> "As a bank we know that we have to go to the customer and not the other way round. Time is money. The future is bright in Libya on the business perspective; tourism will also have a big role."

The Mahari Hotel has been the witness of many events that concluded with important investment agreements.

United Insurance Co.

Mr. Omar K. Labiad
Deputy General Manager
Al Fatih Tower, 2nd Floor, P.O. Box 91809,
Tripoli
Tel: + 218 - 21 - 3351140 / 9
Fax: + 218 - 21 - 3351150 / 1
E-mail: united@muttahida.com

Created in 1999, the United Insurance Company was established to meet the needs of its country, and is currently the number one private insurance company of its kind in Libya. It is the number one company in its sector in terms finance ability, profitability and services. Among the services offered by the company we may find banking services, personal loans and handling claims. The United Insurance Company's key advantages to be underlined are firstly that it offers better services than its main competitor; it is a fully computerised company, using new methods of insurance processing and also the company offers new services to its clients. There are currently 600 private shareholders of the company, six of the major commercial banks holds 5% of the shares each, another 20% of the shares are held by the National Oil Corporation of Libya and the Libyan Iron and Steel Corporation holds another 12 % of the shares. The company has three different branches and six offices including the headquarters located in the capital, Tripoli. The Company represents 50% of the insurance sector of the country, as any investor and contractor must insure themselves or their businesses through a Libyan company the United Insurance Company benefits from most of the oil companies, foreign companies and important local companies. The company has plans to achieve a new project before the end of the year 2004, consisting on a joint venture with a European company in order to provide medical services as well as health insurance. In addition, the United Insurance Company is involved in a four point five billion U.S. dollars Libyan-Egyptian project, insuring most of it.

'We are ready to serve any new client and to guarantee their satisfaction'

TELECOMMUNICATION & IT

OVERVIEW

The country has a modern telecommunications system providing high-quality services between the country's main population centres. Under the Government's five-year development plan up to 2005, emphasis has been placed on the continuing expansion and upgrade of the nation's telecommunications infrastructure.

Domestic Infrastructure

The Libyan government owns and operates the fixed-line and mobile network systems. The postal system is also nationalized with post office box and PTT facilities in all the large towns.

Telecommunications in Libya were greatly improved in the late 1970s and early 1980s. The interior of the country was served by various systems. Radio relay and coaxial cable extended to numerous points and a domestic satellite system was constructed to serve areas not fully integrated into the ground-based networks. The number of telephone lines increased from 90 000 in 1978 to about 215 000 in 1985 with an average of 1 telephone for every 100 citizens. Switching was predominantly automatic.

In 1975 a microwave system connecting radio, telephone, and television signals along the coast was established; it was superseded in 1985 by US$ 25 million high capacity cable system and submarine cable that linked the whole coastal strip with parts of the south all the way to the Chadian border. The transmission systems included microwave radio relay, coaxial cable, submarine cable, tropospheric scatter, and satellites. The system was capable of serving approximately 10 million telephone subscribers, including those along the densely populated Mediterranean coast.

In 1987 Libya had a modern telecommunications system that provided high-quality service between the country's main population centres. The General Post Telephone and Telegraph Organization, a subsidiary of the Secretariat of Communications, carried out all telecommunications activities. Following table shows the telephone lines statistics (2000-2001):

Telephone Line Statistics

Telephone Lines	Year	No./%
Fixed telephone lines	2001	610 000
CAGR (fixed lines)	1995-2001	13.7%
Fixed – line teledensity	2001	11.5%
Total telephone lines	2001	667 200
Total teledensity	2001	12.83%
Digital lines	2000	70%
International outgoingtelephone traffic	2000	43 million minutes
Public payphones	2000	450

Sources: industry and company data

The national teledensity of almost 13% by the end of 2001 is relatively high compared with other African countries. The national target is to achieve a telephone density of 37% by the year 2020.

The majority of households in the capital city, Tripoli, have access to telecommunications facilities, whilst in rural areas teledensity is low, and the installation of main lines costly.

Total demand for telephone lines reached 685 000 by the end of 2000, of which over 88% were covered. Of the total 605 000 fixed lines available in 2000, over 36% (200 000 connections) were in the capital city. Average waiting time for a telephone connection was around 12-16 months.

The total capacity in 2001 is 700 000 lines, the total connected 610 000. (90% of landlines are now a digital network). Average waiting time for a landline connection installation is about 3 months. Table and figure show the fixed line services during 1995-2001:

Fixed telephone lines & Teledensity 1995-2001

Year	Fixed lines	Teledensity
1995	318 000	5.88%
1996	380 000	6.80%
1998	500 000	8.36%
1999	550 000	10.05%
2000	605 000	10.79%
2001	610 000	11.83%

Source: Paul Budde Communication based on ITU data

The latest contracts include the installation of a Siemens digital switch adding a further 8 000 lines in the region of Tubroq to the country's rapidly expanding telecom infrastructure, and the laying down of a submerged fiber optic line (awarded to Alcatel) on the coast from Tripoli to Tubroq, supervised by GPTC.

Fixed Telephone lines 1995-2001

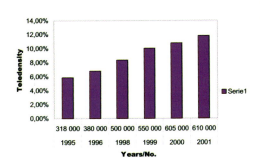

Source: Paul Budde Communication based on ITU data

INTERNATIONAL INFRASTRUCTURE

Submarine cable networks

International telecommunications links, like the domestic routes, were linked via multiple transmission systems. Submarine cables extended from Tripoli to Marseilles, France, and Catania, Italy, provided telephone and telegraph circuits between Libya and Western Europe. A satellite ground-station complex located near Tripoli operated through the Atlantic Ocean and Indian Ocean satellites of the International Telecommunications Satellite (INTELSAT) organization. Additionally, Libya was a member of the regional Arab Satellite (ARABSAT) organization. There is also microwave radio relay to Tunisia and Egypt and tropospheric scatter to Greece.

Under the April 2001 agreement between the African Regional Organisation for Satellite Communications (Rascom), GPTC and Alcatel Espace, Libya is helping to fund Rascom's first satellite under a Mauritius registered company, Rascomstar. Other owners include 41 African telecommunication companies with equity of US$ 41 million. The project will provide Africa with telephone services, Internet access as well as radio and television broadcasts. In addition, the project will provide various telecommunication services to isolated African villages and rural areas currently lacking services. Finally, Af Sat will also provide direct communications between all parts of Africa instead of depending on highly expensive Western links.

In 1998, the GPTC awarded a contract to Alcatel to supply and install Libya's first fibre optic submarine telecommunications system. Known as the Libyan Fibre Optic Network (LFON), the cable stretches over 1 660 km in eleven separate links along Libya's coastline. Completed in 1999, the LFON operates using Synchronous Digital Hierarchy (SDH) over two fibre pairs, which carry voice, data and videos traffic.

125

United Nations Development Program (UNDP) [1]

The General Post Company has signed a number of agreements with UNDP for telecommunications development. The International Telecommunications Union (ITU), in conjunction with the local counterparts, has carried out the projects. These have included:

Formulation of a comprehensive telecommunication plan 1995-2020:

A project carried out between 1993 and 1995, planning for future telecommunications needs, including new technology, improved services and increased phone lines.

Training in telecommunications:

A project implemented between 1997 and 1999, building on the results of the 1995-2020 planning project, which used local and foreign experts to provide training to more than 1,000 employees covering skills improvement and the use of new technology, and finally providing equipment for training centers.

A consultative and maintenance planning study:

This project identified network maintenance needs, and put in place the necessary programs for maintenance and performance improvement; completed.

In 1999 the Libyan General Company for Postal and Telecommunications Services began a project with UNDP and the Universal Postal Union (UPU) for the improvement of the national postal service.

General Posts and Telecommunications Company (GPTC)

The General Posts and Telecommunications Company is the national operator and regulator for all telecommunications services in the country, see table below. As a government organization, the GPTC has no administrative and financial autonomy. The separation of postal and telecommunications functions has not yet been carried out. To date, there is no known government policy for the liberalization and privatization of the telecommunications sector.

Main Telecommunications Organization

Regulatory authority	General Directorate of Posts and Telecommunications (GDPT)
Main public telecommunication operators	General Post Telecommunications Company (GPTC)

In Tripoli your will be surprised by the quality of road transportation and mobile systems network

[1] *UN and Libya, Record of technical co-operation 1950-2000. P 35.*

Mobile Network System

The country's first and only GSM-900 cellular mobile telephone network went live in November 1996 but since then, annual growth has not kept pace with many other African nations. By end 2001 penetration of mobile phones was still less than 1%.

The GSM-900 cellular network was launched with network services offered by Al Madar. Telecom Co. (Orbit Telecom), a subsidiary of GPTC, under an unlimited license covering the coastal zone.

A new mobile phone network has launched its 60,000 mobile lines, bringing competition to the sector for first the time. However, since the new company, Libyana, is state-owned like its rival, some critics question whether the competition will be genuine. Although prices have already fallen from $3,300 in 1997 to $68 plus $410 deposit, this remains much higher than in neighboring countries.

In September 2004 the state-run post and Telecommunications Company signed a 200 million euro ($244 million) mobile phone network agreement with Finland's Nokia and French Telecom equipment producer Alcatel. The deal will involve the creation of new mobile phone network with a capacity of 2.5 million phone lines.[2]

A company source said that the contract worth lies within the scale of 175 million euros, and will be executed over a twenty month period, pointing out that the project network is third generation which will enable customers to have access to the internet, and to send voice and images as well as the full range of Satellite channels.

Under the terms of this contract, Alcatel will supply GPTC with its industry-leading EvoliumTM mobile radio access and core network solution to service 2.5 million GSM/EDGE

and 3G/UMTS users throughout 75% of Libya.

Furthermore, Alcatel will develop and integrate for GPTC a complete portfolio of attractive mobile services to be run throughout the operator's network over the whole country.

Nokia on the other hand will implement West Libyan networks from Tripoli to the western mountains, whereas Alcatel will implement east and south of Libya, the source added.

By the end of 2005, Libya will have taken strident steps in the mobile system, and the mobile average will be 80 mobiles per one hundred people, a percentage regarded as high by international standards, explaining that the target of the company in the field of the mobile system is to cover the entire of Libya.

During his speech at this occasion, Eng. Mohammed Muammar al-Gadhafi, Secretary of the General People's Committee for Posts and Telecommunications, said mobile services would be available even to the most remote villages in Libya.

Five international companies participated in the tender for this project ending up with the selection of these two companies, he said.

Linking Libya through optical and fiber glass and cables with Europe will turn Libya in an important link in the communications between Europe on one hand and East, Central and West Africa on the other.

INTERNET SERVICES

Internet services are reasonably reliable, relatively cheap, and widely available.

[2] *MEC Libya business review 20/09/2004.*

Main Sectors of the Economy:
Telecommunication & It

Internet usage and ISP's

Libya has one main ISP: Telecom & Technology (LTT), connected to Canada's backbone through satellite connection. The bandwidth of this connection is about 5MB. LTT also connected to Italy by fiber optic cable with a capacity of 150MB and a bandwidth of about 45MB.

The main ISP (LTT) has offices (that include access servers) in Tripoli (which is the main branch), Benghazi, Zawia and Misurata. Customers in these cities dial local numbers to connect to the main ISP in Tripoli, where it provides dialup connection with about 200 modems.

The main ISP provides also leased line connection in Tripoli using wireless connection, and DSL cable using PSNT Network connection in Zawia and Misurata.

In Tripoli the dial up connection is respectively good with rates of about 33 kbps, but end users have to endure some problems due to line conditions, mainly humidity and cable disconnection. Large cities have adequate dialup connection status, but in some small towns the line condition prevents using Internet at all. The following table and figure show the number of Internet users:

Internet users 1999-2001

Year	Users
1999	7,000
2000	10,000
2001	20,000

Source: Paul Budde Communication based on industry data

Internet Subscribers Trend

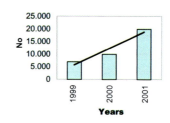

Table below shows Internet statistics for 2001

Internet statistics

Key Internet access provider	GPTC
Internet users 2001	20,000
Internet penetration 2001	0,35%
Internet host computers 2001	70
Internet service providers 2001	3

Source: Industry and company data

Table and figures below show Internet host computers:

Internet host computers 1998-2001

Year	Hosts
1998	4
1999	5
2000	29
2001	70

Source: Paul Budde Communication based on industry data

Host Computers 1998-2001

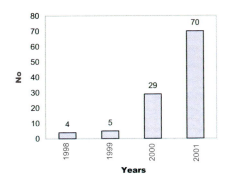

Source: Paul Budde Communication based on industry data

Sub ISP's

There are now four sub ISP's in Libya (BSISP, MWC, CIT and Falk); they are nearly the same size, where each ISP has a link from the main ISP (LTT) via DSL with bandwidth of about 2MB each.

Two of these ISP have one-way satellite connection with 2MB CIR up to 4MB. They have also one-way link from Teleglobe on Asia beam, which requires 3.5m satellite antenna, however using the Asia beam lowers services cost.

Regarding other ISP's, they provide DSL cable connection only in Tripoli and each of them have about 30DSL subscribers, and about 1000 dialup subscribers. (These ISP's do not offer free call option).

Number of Internet subscribers as of Feb 2003

LTT: the number of subscribers along it's four branches (Tripoli, Misurata, Zawia and Benghazi) is about 15 000 around the country, more than 10 000 subscribers in Tripoli, each branch has access servers for dial up connection, subscribers in other regions can access internet by dia-

ling one of the branches using the same account, this ISP provide DSL connection using wireless connection in Tripoli, and using cable connection in Misurata and Zawia, DSL is used only by some companies and few number of internet café's.

Bait Alshams: offers both dial up connection and DSL in both Tripoli and Misurata, The number of dial up subscribers is about 1200 users, in addition to a few number companies and internet café's (about 30 internet café's) using leased line (DSL), it's the only ISP that offers flat rate accounts (monthly payment for unlimited hours), the demand for this type of accounts is good for internet café's, where customer can get it easily at any time from any of the company agents.

Modern World Communications: about 2000 dial up subscribers, and about 30 DSL subscribers, it's the only ISP that offers prepaid cards, there is high demand for these cards.

Alfalk: A new ISP (2002) offers only dial up connection, the number of subscribers is about 500 subscriber, this ISP allows it's agents to activate accounts online through it's web page and the customers don't have to pay activation fees, this made the subscription easier and caused good demand for it's services.

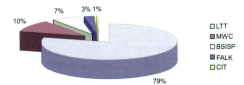

Internet Via Satellite Services

CIT: this ISP offers only dial up connection, the number of subscribers is less then 100, where it encounters many technical problems. Figure above illustrates the subscriptions/ISP distributed in percentages.

Note: the number of Internet subscribers differs from the number of Internet users, where more than one user may use the same account. The hourly price is LD 0.700 /hr.

Most of the demand is for one-way services. The two-way service requires a special usage permit, and only some governmental companies have the permit to use two-way services. Oil companies need it for their oil fields that are located in the desert to provide connectivity to their networks using VPN. Some companies connecting network are using DSL for adequate distances, and using dialup for larger distances. There is high demand for one way services for Internet cafes, where there is a large number of Internet café's using dialup connection supported with satellite connection, and some of them using DSL. Integrated service digital network (ISDN) access is not available yet.

Internet café's can be classified as follows:
– Small size: which contains about 6-10 PC's and work for 8 hours daily.
– Medium size: which represents the majority and contains about 10-15 PC's and work for 15 hours daily.
– Large size: Which contains more than 20 PC's and work about 24 hours, table below.

Average monthly subscription fees as per class of Internet cafe
(Internet via satellite services)

Class of Internet Café	Speed required	Estimated Monthly Price US$
S	64-128	60-80
M	128-256	80-120
L	256-512	120-200

One-way Internet via satellite:
– Available on Arab sat, Nile sat, and Eutelsat.

Services available:
– Static IP services: the best and most wanted.
– Proxy services: good for browsing and downloading (Uploading is not available).
– VPN service

The customers prefer unlimited services even if the speed is slower, and prefer also a stable slow service than high-speed service with frequent disconnections.

Prices of Internet services via satellite:

Arab Sat 2B30
1) Speed 256K-Tilleque LD 900 quarterly.
2) Speed 512K-Tilleque LD 1725 quarterly.
3) Speed 1024K-Tilleque LD 1650 quarterly.
4) Speed 256K-Static IP LD 1075 quarterly.
5) Speed 384K-Static IP LD 1325 quarterly.
6) Speed 512K-Static IP LD 1650 quarterly.

Sesat Sat
1) Speed 128K-Tilleque LD 360 quarterly.
2) Speed 256K-Tilleque LD 720 quarterly.
3) Speed 512K-Tilleque LD 1440 quarterly.

- The subscription fee amounts to LD 200 paid once when signing the contract.
- The minimum period of subscription is 3 months.

MAIN COMPANIES

Alcatel

Mohamed L. Bala
General Director
Dat El-Imad Tower 5, Floor 15
Tripoli
Tel: + 218 - 21 - 3350130 / 1
Fax: + 218 - 21 - 3350133
E-mail: Mohamed.I.bala@alcatel.fr
Website: www.alcatel.com

Alcatel provides end-to-end communications solutions, enabling carriers, service providers and enterprises to deliver content to any type of user, anywhere in the world. Leveraging its long-term leadership in telecommunications network equipment as well as its expertise in applications and network services, Alcatel enables its customers to focus on optimising their service offerings and revenue streams. Alcatel is organised along markets with three business groups: the Fixed Communications Group, the Mobile Communications Group and the Private Communications Group. Alcatel had sales of 12.513 billion euros in 2003 and currently operates in more than 130 countries. Alcatel is a key player in African telecommunications. The company is responsible for building the first satellite that will exclusively cover the continent for both telecommunications and TV broadcast. This project, the "Rascon Sat", is the first to be created by a consortium of African countries and it will not only reach new rural areas but will also allow the communication between African countries without having to use other networks outside the continent. 25 years ago, Alcatel operated a microwave analogue transition project in the southern part of Libya and as a result a good relationship between the French company and Libya started. In December 2002, Alcatel was selected by AGIP GAS B.V., a joint venture between AGIP North Africa and the National Oil Corporation of Libya (NOC) to provide a full turn-key tele-communications network and associated services for AGIP GAS's internal communication needs. The contract, worth 58 million euros, fits within the framework of the global Western Libya Gas Project (WLGP) aimed at developing and exporting large volumes of natural gas via pipelines to Europe. Alcatel was chosen for designing, building and commissioning a sophisticated communication network consisting of telecom and non-telecom subsystems such as access control, closed circuit television and meteorological systems. In 2004, the state-run Libyan National Company for Post and Telecommunications awarded a 200-million euro deal to build a mobile phone network joint project to Alcatel and Nokia. This project involves building a mobile telephone network with a capacity of 2.5 million lines throughout Libya; 1.5 million of which are corresponding to the French company. In addition, the company was awarded a project for implementing the fibre-optic backbone that connects 13 different towns to each other in the north of Libya. This system may soon be connected to Tunisia and possibly Egypt in the near future. These important projects are an example of Alcatel's success in Libya. The company has a long history and is fully trusted and has a good team and a good strategy towards the future. Libya is opening up; all the companies are coming back. There are changes in policies and there are good possibilities for Alcatel in Libya for the years to come.

"Newcomers have to come to Libya with an open mind; it is difficult to succeed with European habits. The country is changing and the people are opening their minds too but things cannot be done that fast."

"I HAVE
MY OFFICE WITH ME
WHEREVER I GO."

Alghad IT Services

Mrs. Jamila M. Durman
General Director
Aswae Al Ketoony St. P.O. Box 91590, Tripoli
Tel: + 218 - 21 - 3341503
Fax: + 218 - 21 - 3341503
E-mail: alghad@lttnet.net

Alghad IT Services first started its operations in the year 2000 as the IT sector was developing new opportunities and needs. The company is always implementing the quality of its services in terms of diversification and expansion, offering both software and hardware as well as networking and training solutions. When it comes to software, the company is both a reseller for the top quality brands as well as a designer of custom made computer programs meeting any needs that a company or individual might require. Training of the staff that will be using the custom made programs can be done by Alghad IT services at the modern and well-equipped offices of the company located in Tripoli. The need of IT solutions in Libya is promising and looks to have huge investment potential. At the present time there are few companies in that field, which means that the company might benefit from a huge market share presently and in the years to come. Currently, the management of the company is working on a joint venture with a European company in order to provide technical assistance to the latest IT products handled and resold by the company. This same management is also carrying out a joint venture with a plant assembly for similar matters. IT business is essential and important nowadays. Libya needs to go side by side with international advances that are made in the IT field, as IT is the common and global language.

"If everything goes well, IT will soon become a booming sector in Libya and we will achieve success within it."

Al-Madar Telecom

Musa S.H. El-Naas
General Manager
P.O. Box 83792
Tripoli
Tel: + 218 - 21 - 3609089
Fax: + 218 - 91 - 2120161

Al-Madar was created seven years ago and it now operates three hundred lines in Libya. With close to three thousand employees, Al-Madar is 95% publicly owned and has a yearly income of approximately one hundred and fifty million Libyan dinars. During 2004, the network of the company was quite limited and was operating at 95% capacity. However, since November 2004, serious improvements have been made. Due to the size of its network, the company still maintains quite high pricing in the initial purchase of the Al-Madar GSM cards. This will change according to the planned growth of the company in the years to come. Currently, the advantages offered by Al-Madar are an exclusive knowledge of the telecommunications industry in Libya and a unique position with the G.P.T.C. (organism that regulates and controls the telecommunication sector and the postal services in the country) due to their close collaboration since the beginning.

TNT (ADM for Air Cargo Services)

Kareem Ezmerli
WEF Administrator
Gurgi near Al-Andalus Market, P.O. Box 81206, Tripoli
Tel: + 218 - 21 - 4770326
Fax: + 218 - 21 - 4778009
E-mail: admlyexpress@yahoo.com
Website: www.tnt.com

TNT is one of the largest companies for express services with many years experience in this field. TNT is specialized in delivery, documentation and cargo. TNT ADM International Express Services was an exclusive agent and partner of Thomas National Transport International Express (TNT) established in 1946 and became fully owned in the name of Mr. Mohammed D. Fazzani on April 21, 1993 and now operates with a total of eight branches within Libya. TNT Express now operates in over 222 countries and delivers 3.6 million parcels, documents and pieces of freight a week using its network of nearly 1,000 depots, hubs and sorters centres. TNT Express carried over 187 million consignments in 2002 with a total weight of over 3.3 million tonnes. TNT Express operates 43 aircraft and approximately 21,000 road vehicles. TNT ADM International Express Services includes door-to-door collection and delivery services. TNT provides time or definite delivery and packing. Whatever the clients want, TNT does it, cargo, by land and by sea and full track and trace visibility. All the shipments are traced with a high technology computerized network, which enables it to solve all potential problems with door-to-door and airway bills deliveries as soon as possible.

"TNT offers its distinguished experience with unbeatable prices. We aim to offer all our clients efficient and effective service."

INDUSTRY & TRADE

Trucks & Buses Co.

For Further information please contact us at:
Tel: +218 21 370 5034 / 370 5039 / 370 5040
Fax: +218 21 370 5036 P.O.Box: 30768
info@tbco-libya.com www.tbco-libya.com
Tajura, Tripoli - Libya

OVERVIEW

During the long embargo period, Libya had to become self-sufficient. During that time many companies were created that had to fill the industrial needs of the country. Companies such as Trucks & Buses or Aman Company for tyres (formerly Tyres & Batteries) have experience as the only industrial suppliers for their specific products in the Libyan market. Both of these companies are located in the vast Tajura Industrial Complex near Tripoli. Now things have changed. With the opening up to the world markets, Libya not only wants to offer opportunities in the oil and gas industry, but also in other fields related to light and heavy industry, as well as the very promising trade sector. The government, through the General Board of Ownership, Transfer of Public Companies and Economic Units, is undertaking a very ambitious privatization process of those companies that have led the industrial sector in the country.

The G.B.O.T.

The General Board of Ownership, Transfer of Public Companies and Economic Units (GBOT) was established by the resolution of the General People's Committee No. 198 in 2000. It reports to the General People's Committee and is delegated to execute the program of transferring public company and economic unit ownership to the private sector. In doing such, it widens the base of ownership to accomplish people's socialism. This is done in turn to give a bigger role to domestic activity in the economic sector, to develop proficiency and competitive capability in the production and services sector and to contribute in the development process.

Agriculture

Although it contributed only 9% of GDP in 2002, agriculture is considered a major sector in the national economy. In 2001 the number of farm owners was 176,658, 33.7% of whom were practicing agricultural work as their main

Headquarters of Aman Company for Tyres in Tajura, Tripoli

activity (only 59,566 people in total). When comparing the results of the two agricultural censuses that were made in the years 2001 and 1987, one can see that most of the new people engaged in farm work were not actually farmers, but other people who wanted to have a second activity. Furthermore, the arrival of water from the Great Man-made River Project to those areas encouraged many people to buy agricultural land for the same reason. Nevertheless, it is worth mentioning that the number of farmers who were dedicated to agricultural work was less in the year 2001 in comparison to the 1987 census with a ratio of 38.3% to 33.7% while the ratio of the farm owners who were not dedicated to agricultural work increased. We see that the agricultural activity is concentrated in certain areas in the north and south of Libya. The total sum of the area of agricultural land in Libya is nearly equal to 1,809,596 hectares. As a result of the dry weather and water shortages over the last years, 62.2% of those areas did not receive water.

The total number of agricultural workers who were dedicated to work on agricultural farms according to the results of the economic and social survey carried out between 2002 and 2003 was 65,073 farmers. This was thus a ratio of 4.8% to the total manpower, which had a total of

1,364,709 people. Over the last ten years, Libya has witnessed a decrease in the number of most fruit trees. This was due to the increase in the number of farm owners who were not dedicated to agricultural work in combination with the scarcity of water and the urban extension occurring on planted areas neighbouring to the cities. Such a decrease in trees, however, did not include the date palm trees whose number increased to 3,439,885 according to an agricultural census in 2001. This represented an increase by a ratio of 1.6% to the 1987 census results. As for the total number of other fruit trees: olive trees represent 26.77% of the total number of the fruit trees in Libya with a number of 5,679,964, date palm trees represent 16.47%, almond trees 11.45%, grape trees 16.30%, citrus trees 12.41% and fig trees represent 4.90% of the total fruitful trees. In contrast, the number of the animals increased during the same period, as the number of the sheep recorded was 5,642,895 with an increase of 29.3%. In this regard, it is worth mentioning that Libya witnessed the reclamation of tens of thousands of hectares of lands to be used for agricultural projects distributed to people during the 1970s. Since Libya was suffering from a shortage in the water resources, it was necessary to start the implementation of the Great Man-made River Project. The water from the project will contribute substantially to development in the agricultural sector over the coming years.

The Free Zones

Law no. 9 for the year 2001 concerning the organization of transit trade and free zones and its executive regulation, which include the following points, regulates the creation of free zones in Libya:

Purposes

– Transit trade promotion and preparation of goods for commercial exchange purposes and market requirements.
– Encouraging all kinds of manufacturing means and transformable operations.

– Technology and know-how transfer and settlement, and developing it by freeing it from restrictions.
– Opening of work scope and local subjects training.
– Releasing the joint interest for both national economy and investors.

Areas

– Establishing plants, offices, stockrooms, warehouses, utilities, selling halls, installations, transport materials, communications and all facilities required for the purpose of investment or utilization.
– Utilization of real-estate, use and exploitation of facilities and installation in the zone.
– Providing assistant services such as banking and insurances services, communication and transport services etc.

Merits and Guarantees

– Relieving projects and investors in the free zone from registration in import and export records and trade register.
– Exemption of projects from released incomes and all types of taxes and fees.
– Relieving acts, bills, assets, money exchange and transactions in the free zone from taxes and fess.
– Relieving transit goods and others goods, merchandise, services, properties and exchanges coming or leaving or exchanged in the zone from any duties or fees except for service fees.
– No seizure of projects and investor assets on such rights unless by legal text or judicial procedure.
– No project nationalization, or confiscation, or take over, or expropriation or subjecting it to procedures with the same effects except legally and with a just compensation without discrimination.
– Permission of partial or full projects property transfer to another investor.

Engineering Industries Co.
For further information please contact us at:
Tel: +218 21 370 5481   Fax: +218 21 5415
P.O.Box: 30485   Tajura, Tripoli - Libya

ENGINEERING INDUSTRIES CO.

At Industrial Engineering Company, we are specialized in galvanized steal constructions, electric towers, gas cylinders, LPG cylinders, and hangars, handling an average of 5.000 to 20.000 metric tons of steel per year.

Our prices are very competitive due to cheap costs on energy, labour and raw materials that can be found in Libya.

Industrial Engineering Company is an expanding and dynamic Libyan company, planning to expand exporting up to 50% of the production to foreign markets in the near future by offering quality products, and continue being the biggest company in the sector.

The Great Man-made River Project

The Great Man-made River Project (GMRP) is one of the largest civil engineering projects in our time and the longest water conveying scheme ever. The project is aiming to extract 6.5 mcm/d (six and half million cubic meters of water per day) from the southern Libyan dessert and convey it to the coastal fertile land in the north where more than 85% of population resides. The project will extract the required flow from approximately 1,300 wells that are distributed over an area of about 17,500 square kilometres. The depth of the wells ranges from 450 meters to 1000 meters. The water will be collected in collecting network (collectors) which extend over 2,250 kilometres of pipeline having diameters ranging from 400 mm to 2800 mm. The water will be conveyed in a long conveying network, the total conveyance length of which is more than 4,000 kilometres. The conveyance is composed of 575,000 pipe joints. The pipes' diameter varies from 1600 mm to 4000 mm, and the pressure ranges from 6 bars to 26 bars depending on the topography of the terrain. The length of the pipe is 7.5 meters and it is made of pre-stressed concrete cylinder sections (pccp).

The project is managed by The Implementation and Management Authority of The Great Man-made Project, which was established in 1983.

The overall project is constituted of basically three phases as follows:

Phase 1, which is called the Sarir Sirt–Tazerbo Benghazi System, is to convey 2 mcm/d and has a conveyance of 1600 kilometres of 4 meter pipeline. It conveys water from Tazerbo and Sarir well fields to Benghazi and Sirt cities and all the towns in between.

Phase 2, which is known as Hasuana- Jafara system is to extract and convey 2.5 mcm/d and has conveyance pipeline of 1000 kilometres of 4 meters pipe diameter. It conveys water from Alhasuana well fields to the Jafara plain, the city of Tripoli and the other towns on the way.

Phase 3 is complementing phase 1 and phase 2 and is composed of the Alqordhabia Assadada Link which links phase 1 with phase 2 and is capable of conveying 98,0000 cm/d in either direction. It is a 4 meter pipeline conveyance and is 190 kilometres long. The Kofra system is to extract 1.68 mcm/d from the Kofra well field and convey it to a Sarir header tanks location to join the phase 1 network and increase the daily flow to 3.68 mcm/d. The length of the pipeline is 480 kilometres of 4 and 3.6 meter diameter pccp pipe line. The Ghadamis-Zuara-Alzawia system is to exploit and convey a daily flow of 250,000 cm/d from Ghadamis well field to the western coastal towns west of Alzawia city. The conveyance is the 650 kilometres of 1600 mm pccp pipeline in the Aljaghbob-Tubroq system is to exploit and convey a daily flow of 135,000 cm/d from a well field south of Aljaghbob to Tubroq city and eastern towns around it. This system will also cover the daily water demands for all the towns from Derna to Albordi. The conveying network is about 600 kilometres of 1200 mm pipeline. The project will be using 16 pump stations to supplement the gravity flow, which characterizes the GMRP systems, and it will consume about 500 MW of power.

The GMRP is providing an annual consumption of over 1000 cubic meter of water per capita and will be distributed to users as follows:

Municipal use	30%
Agricultural use	68%
Industrial use	2%

Its aim is to irrigate approximately 135,000 hectares of cultivated land and to provide most of coastal cities and towns with their domestic water needs.

MAIN COMPANIES

African Sail Shipping Company

Capt. Salah El Jamal
General Manager
El-Fateh Tower 2nd floor No.3, P.O. Box 93043, Tripoli
Tel: + 218 - 21 - 3351387 / 3351388
Fax: + 218 - 21 - 3351390
E-mail: assa2000-ly@lttnet.net
Website: www.assa-agent.com

The African Sail Shipping Company started its operations in 2000 as the first Libyan private shipping company. The company was a pioneer among the 140 to 155 companies now operating in the field and managed its first steps very well due to its good relationship with the international agents operating all over Europe. The company operates "door to door" all along the sea coast that borders the northern part of Libya bringing in all the equipment or goods that have to be transported from any point overseas, the same from Libya to any point of the world. The company takes care of the paperwork, customs, immigration and port authority requirements if needed. The main competitive advantage of this company is the quality of the service provided, and the trust that the international agents have for its management. Trust is the most important factor to take into account when it comes to the transportation of goods or equipment through other people as when the shipment reaches the country, they act as owners of the shipment. The company not only has experience dealing with European countries such as Italy, Germany, France, but also with Canada and with the Far East. It is the only shipping company in Libya registered in the "Binco" international shipping agencies listing. The company now needs investment to grow and despite the difficulties coming from the current legislation it needs to expand to more services and facilities. Currently the management is looking forward to finalising negotiation sessions with major international firms in order to operate jointly on the Libyan ground.

"When shipping anything to Libya with us all the papers should be correctly sent before."

Akida Co.

Dr. Abdunnaser Akak
Chairman
Ajraba St. 55 Zawiet El Dhmani
Tripoliz
Tel: + 218 21 3406250
Fax: + 218 21 3406254
E-mail: akidalibya@mail.com
Website: www.akidagroup.com

Akida Group is a consortium of Libyan companies established in 1988 by a group of experts in field of import and export of consumer electronics, electric home appliances, and food stuff as well as building and construction materials. With up to 220 employees, the group headquarter is located in Tripoli, Libya. During years of trade activities the group has achieved superb business relationships with many leading international companies. As well, Akida gained significant market share, by applying modern management methods in marketing. The company provides high quality services to clients through a chain highly organized distribution network and the group owns four hundred distribution outlets all over Libya.

Our slogan is "together for a better future". We are open and prepared for participation in partnership to realize it.

RECOMMENDED PARTNER

Aman Company for Tires

Eng. Sulieman Abu-Sa'a
Chairman
Tajura – Tripoli
Tel: + 218 - 21 - 3705090
Fax: + 218 - 21 - 3705117

The Aman Tires Factory used to be part of the Aman Company for Tires and Batteries created in 1976. In 2004, after a reorganisation process which split the first company into two new others, one for tires and the other for batteries, the Aman Tires factory was created, better adapting to the right framework that would allow welcoming new investments. In 2000, the mother-company for tires and batteries used to have 1,300 employees and had many branches outside Tripoli selling its products. The Aman Tires Factory kept 900 of that workforce and currently the new companies has cancelled those branches in order to focus primarily on production and direct sales. Some of the competitive advantages of this Libyan Tires factory are the high quality of the product compared to other international producers and its cost and the benefits of operating in Libya. Firstly, the product quality reaches the DNA technical specifications and the production plant uses both top quality machinery and highly trained staff. Secondly, the production cost is cheaper than other places considering the price/quality ratio as with, for example, the synthetic rubber coming from oil is very cheap. Libyans used to find Asian products cheaper than Libyan ones but not offering the same quality. Now, the company, which has a market share of approximately 30%, is seeing large increases in this same market as people realise that the quality/price ratio offered by the company is much more interesting. This will also happen in the future if the country creates better quality control over the goods imported for overseas. Finally, Libya itself offers many advantages, starting with the lack of religious conflicts as all Libyans have strong ties among themselves. Libya is also well situated to the important sur-

"High Technology equipment used at the Aman Company for Tires"

rounding markets (Europe, Africa).

"I invite everyone to visit Libya, maybe we need to improve our behaviour in some cases as well as our organisation, but the fact is that you can freely discuss everything here in Libya, a difference it shares with any other democracies. On the economic perspective, we have and we need more industry, from food products to all kinds."

Arab Union Contracting Company

Eng. Ziad Adham Al Muntasser
Chairman
Al-Fateh Tower, Floor 22, P. O. Box 3475
Tripoli
Tel: + 218 - 21 - 3351201 / 2
Fax: + 218 - 21 - 3351203 / 5

The Arab Union Constructing Company is specialised in construction and contracting works. An example is that the company built the famous Al-Fateh Tower in Tripoli. The company was created as a joint Arab venture by the former federal governments of Libya, Syria and Egypt. Currently only Libya and Syria are in the joint venture. Its operations started in 1976 with the construction of schools, hospitals, and plants among other things. The head office of the company is located in Tripoli with a branch in Damascus, Syria. In Libya the company not only owns the Al-Fateh Towers, but it is very much involved with the big housing project that is going on in the country. This housing project, undertaken by the Libyan authorities, consists of building an initial 60,000 housing units this year and reach in a middle and long term basis an average of 50,000 units per year in the years to come. Managing Al-Fateh Tower as an own investment, the company rents offices to some of the most

important international and local companies settled in Libya at a rate of US$ 30 per square meter on a monthly basis. In addition, and due to the shortage of cement in Libya, the company has been operating its own cement factory since March 2004, located 150 km from Tripoli. This first Libyan privately owned cement factory will have an initial capacity of 1.4 million tons per year, fulfilling the company's construction needs. The company also manufactures its own tiles and press beams and will become the biggest producers of ready mixed concrete with a production of more than 1000 cm? per day. Regarding the future, the management has great expectations for the company. It has key competitive advantages like experience and equipment, which are strategic in a country like Libya. There is also a big shortage of tourism resorts and hotels so it is already proven that there is a large potential within the construction sector in Libya for the years to come.

"Come now before it is too late; all the investors were shy before but now that the embargo is over there are a lot of foreign delegations coming, even from the U.S. As we say, the first to come will be the first to be served!"

Ashoula Battery Co. Ltd.

Eng. Almbruk G. Wali
General Manager
Tajoura Industrial Area, Tripoli
Tel: + 218 - 21 - 3705101
Fax: + 218 - 21 - 3705112

The company has been operating on its own since the 1st of May 2004 as a result of the division of the Amman Company for Tires and Batteries into two new companies. This decision was taken to benefit those two companies both technically and structurally. The company is the only producer in Libya, producing both 40 to 60 A and 70 to 200 A battery models. The company has 324 employees and is currently going through a privatisation process. The company produces and sells an average of 150,000 liquid batteries per year, covering around 25% of the Libyan market. The main task of the management is to find investment in order to update and increase the capacity of the production plant; that would allow covering the remaining 75% of the Libyan market. This could be easily done as the company offers a high quality product at a very competitive price. Currently there are other batteries available in the country that are taking up the part of the market the company cannot cover due to the limited producing capacity of its plant, but the fact is that those other imported products do not often reach the minimum quality standards. That is due to the fact that the country has opened its market to foreign imports of goods and equipment but has not yet been able to establish a regulatory procedure to verify the quality of those imported products. Among the competitive advantages offered by this company, investors should consider that it is the only producer in a market that has a much bigger demand than its current producing capacity; that the producing facilities are already installed as well as the storage facilities plus the fact that the energy cost in Libya is among the most competitive in the international scheme. The company is also working on a project regarding the recycling of the liquid batteries in order to reduce pollution and another project of diversification of its production in order to build industrial batteries. The company hopes to see investment soon in order to increase its capacity and market share in Libya. Also the management is willing to get the international quality seals in order to reinforce the confidence of future clients and collaborators.

"Libya should soon carry out the control of the quality specifications. We are ready to discuss any collaboration and we hope to get investment soon in order to improve our product as well as increase its capacity and market share."

Electronics General Company

Eng. Ali M. Elmesallaty
General Director
Tajoura industrial complex road
P.O. Box 12580 Tripoli
Tel: + 218 - 21 - 3705231 / 8
Fax: + 218 - 21 - 3705252
Email: eng_prod@garyonis.com

The company was first created in 1976 as a trading company, importing electric and electronic goods from abroad; in the mid-eighties the company started its own production lines. The Electronics General Company is the only producer in consumer electronics in Libya, producing televisions, radios, telephones, components; professional electronics such as walkie-talkies or radio stations, cables and speakers among many other products. Among its products we can find the Gariunes brand for televisions, which is known in Egypt for example. The Electronics General Company was also the first Libyan company to get the ISO 9001 certification in the year 2000. The company is also willing to provide consultancy on electronics projects, production and maintenance. With a total of two thousand employees, the company has twelve production facilities distributed in six different complexes; each of them specialised in a different product. On the social aspect, the Electronics General Company is a pioneer in Libya. Apart from being the first company where the workers were also shareholders of the company, most of the workforce is female. In addition, the company has plans to finalise agreements with the G.P.C. for Education in order to supply up to four thousand schools with computer equipment in order to contribute to the development of the Libyan youth. The company is currently going through a privatisation process and has received many offers from well-known foreign brands to reach collaboration agreements. The management is happy and ready to welcome new investors, the management also sees the privatisation process as something that will diversify the sharing of the company with other people, expanding it and being positive for its development.

"Welcome to foreigners willing to carry out long term investments, they will find comfort, security and friendly people here in Libya!"

Engineering Industries Company

Eng. Abdulla Yahya
Chairman
Tajura P.O. Box 30485, Tripoli
Tel: + 218 - 21 - 3705481
Fax: + 218 - 21 - 3705415
E-mail: eiclibya@hotmail.com

The Industrial Engineering Company, known before as

Industrial Engineering Arab Company, was founded in 1994 with a capital of 165,000 LD. Located in the Tajura industrial complex of Tripoli, it is specialised in several galvanised steal constructions handling an average of 5,000 to 20,000 metric tons of steel per year. The main products of the company are transmission line towers, lighting poles and steel structures such as prefabricated steel warehouses and hangars; the company has an approximate turnover of 30 million LD per year. The management is also planning to develop new products in order to diversify the production and the projects handled. An example is a project to manufacture lighter gas tubes than the ones being used in Libyan houses. Also, there is another project to work within the heavy infrastructure field in projects for houses, bridges, and there are even plans for sports stadiums. Currently, the company is doing the construction of most of the benzene station towers in the country. The main advantage of working with this kind of product is that steel

"Hard work on steel projects at the Industrial Engineering Company"

AMAN COMPANY FOR TIRES

PRODUCING TOP QUALITY TIRES FOR TOP PERFORMANCE

Tel: +218 21 370 5090 Fax: +218 21 370 5117   Tajura Tripoli - Libya

constructions are safe and they can be constructed much faster. In addition, the prices offered by the Industrial Engineering Company are very competitive due to the cheap costs on energy, labour and raw materials that we can find in Libya. The company is planning to expand, exporting up to 50% of the production to foreign markets in the near future. The Industrial Engineering Company is an expanding and dynamic Libyan company, offering quality products. It is the biggest company in its sector.

"Libya is the preferred investment destination in North Africa"

General Tobacco Company

Dr. Salah A. Naily
Chairman
P.O. Box 696 Tripoli
Tel: + 218 - 21 - 4801404 / 6
Fax: + 218 - 21 - 4801405
E-mail: gtc@gtclibya.com

The General Tobacco Company is one of the public companies owned by The General Industrialization Corporation. It was established in 1972 with a capital of 5 million Libyan Dinars. The company specialises in tobacco harvesting, making and marketing of cigarettes, cigars and other tobacco products in addition to matches. The company is run by a people's committee, which heads seven departments, nine offices, one branch and ten sales offices throughout the Great Jamahiriya. The company headquarters are located in Tripoli within an installation of around 57 square acres and a total of 1500 employees. Recently, the company has achieved an agreement with British American Tobacco to share its installations in Tripoli. The company has the monopoly of tobacco producing and importing in Libya. It is also dealing with foreign countries such as Greece, Switzerland or Egypt. Investors should know that General Tobacco Company controls 100% of the industry in Libya. The tobacco industry is a very profitable one, it is a secure investment in Libya and there are big

expectations for the future. In the years to come the company wants to replace and update some of the machinery being used and is ready to welcome foreign investment as it is under a privatisation process.

"Libya is safe, there are no major problems here, the energy costs are cheap, the political atmosphere is strong and stable and we encourage foreign investment!"

National Development & Building Materials Industry Company

Eng. R. El Mahmoudi
Chairman
P.O. Box: 12950 Tripoli
Tel: + 218 - 21 - 4813962 / 64
Fax: + 218 - 21 - 4813965
E-mail: ncdico1@yahoo.com

The company was established in year 2000 among a group of five companies by The National Investment Company, one of the largest investment companies in Libya. Established initially with a capital of one million Libyan dinars, the capital was increased later on to five million Libyan dinars. The activities of the company are the manufacturing and marketing of all kinds of building materials, executing and operating factories related to building materials. Other activities include importing and exporting all kinds of building materials, purchasing licenses and know-how related to building materials industries. It also deals with land purchases and constructions needed for its activities. The company is currently working on a joint venture project called " Hollow Clay Bricks Project " in the Gherian area (about 80 km west of Tripoli). It has a brick production capacity of 130,000 tonnes per year. It is important to know that the local market needs about one million tonnes per year. The company is in the final stage of evaluation of some offers prepared by specialised companies in this field from many different countries. The management is also planning to establish other projects in the future related to the building materials industry such as tiles, marble, sand lime bricks, cement, gypsum, water proof membranes and so on. The Libyan market is a virgin market, which is currently encouraging many investors to take part in it. This is why the company is also seeking to make it clear to foreign investment companies which are interested in the Libyan market that the most successful projects in Libya are ones related to the building materials projects. This is due to the availability of the raw materials in the country in huge quantities as mentioned above.

"We would like to see our company a very active company in its field in the near future, In Shaa Alah!"

National Trailer Company

Eng. Mohamed M. Khaddar
Chairman & General Manager
P.O. Box 12320 Addahra, Tripoli
Tel: + 218 - 21 - 3705061 / 3
Fax: + 218 - 21 - 3705060
E-mail: maktorat@hotmail.com

The National Trailer Company was founded in 1982 as a joint venture between a Libyan truck parts company, which holds 75% of the shares and the Italian company Calabrese which holds the remaining 25%. The company has operated assembling parts since 1984 as a first stage during the building of its workshop, which was completed in 1990. The company produces each year an average of 1000 pieces of all types of track equipment, including refrigeration tanks, tractors as well as semi-tractors. It is providing all the garbage collecting tractors to the municipalities. Also, the company has clients such as foreign oil companies or the famous Libyan Great Man Made River Project. Currently, as the local demand exceeds the production, the

management is looking forward to expanding the company welcoming possible investors. The main advantage for the company in Libya is that raw materials are cheaper, for example the steel which is provided to them from Misurata, as well as the proximity to Europe. Libya offers great advantages as well; it is a peaceful country and the private sector will improve more and more.

"We are in the situation to realise major changes, we need other's experiences and help. We have a very good foundation and land in order to accept the changes for the best."

Overseas Shipping Company Libya

Mr. Omar Dahmani
Chairman
Dat El Imad Tower 5, 1st floor, Tripoli
Tel: + 218 - 21 - 3350870 / 1
Fax: + 218 - 21 - 3350322
E-mail: oscl@lttnet.net

The Overseas Shipping Company of Libya is the agent for the first French carrier CMA-CGM throughout Libya. The company has offices in Tripoli, Benghazi, El-Khoms, Misurata as well as in all the national ports and major oil industry entry points such as Ras Lanouf or Brega. The company, 100% privately owned, was established on the 1st October 2003 as a private agent following the new legal framework and has been an agent of the French company for 12 years now. The company staff has previous experience in shipping that dates back to the sixties. There are 18 employees in Tripoli, 12 in Misurata, 6 in El-Khoms and the remaining at the other points. There is a weekly stopover in all the destinations mentioned above, for all the destinations available. The Overseas Shipping Company of Libya benefits from being the agent of the 1st European shipping company and the 4th worldwide. The French company CMA-CGM works around the world, from the Middle

Arab Union Contracting Company

E FUTURE OF CONSTRUCTION IN NORTH AFRICA, 1.4 MILLION TONS
R YEAR TO BECOME THE BIGGEST PRODUCER OF READY MIXED
NCRETE IN LIBYA.

East to Europe and America, offering services at all the seaports. This enables all the traders and companies to approach any other market easily. The policy of the company is simple but strong. They claim that their better service always provides better results for their clients.

"Libya is a very welcoming country, where the humanity and all of its traditions are highly respected"

Tripoli International Fair

Eng. Khaled Saleh Senussi
Chairman of the Board
Omar Mokhtar St., P. O. Box 891
Tripoli
Tel.: + 218 - 21 - 4440655 / 3332255
Fax: + 218 - 21 - 4448385 / 3336175
E-mail: info@TripoliFair.org
Website: www.tripolifair.org

The Tripoli International Fair is situated on Omar Mokhtar Street. On an area of land of more than 90,000 square meters, 40,000 are covered stands and the remainder is open area. The Tripoli International Fair is considered to be one of the active members of the Union of International

"The Trucks & Buses Company manufacturing plant in Tajura"

Fairs, which is located in Paris. The Spring Period of Tripoli International Fair is held from April 2 to 12. Participation averages around 30 countries and more than 2000 companies and establishments. The fair is an international, industrial, agricultural and commercial event. It observes the regulations and international fair norms. The fair has three departments: 1-The Technical department. 2 -The Exhibition Department. 3 -The Managerial and Finance Department. During the period of the fair there are many festivals in which Libyan folk bands and other international bands participate accompanied by music and singing. The fair also includes a funfair for children's use. The fair, in addition to different stands, also includes an industrial area in which heavy equipment and machinery are displayed. The Tripoli International Fair also offers comprehensive services to international exhibitors and guests of Jamahiriya throughout the period of the fair from 2 to 12 April. The main services provided by the Tripoli International Fair are: electricity, water, hotel reservations, visa invitations, media advertising and assistance in contacting Libyan companies.

We welcome investors to Libya to participate in the development of our country. We are ready to cooperate with all countries world-wide.

Trucks & Buses Co.

Eng. Nasardeen Elhawat
General Manager
Tajura Km. 19, Sidi A. Karim St, P.O. Box
12869 Aldahra, Tripoli
Tel: + 218 - 21 - 3705034 / 39
Fax: + 218 - 21 - 3705036
E-mail: nasreddin@tbco-libya.com

The Trucks and Buses Company was established in 1984 as a joint venture between the General People's Committee of Industry; owning 75% of the total shares and the Italian company Iveco owning the remaining 25%. The company assembles heavy truck components as well as bus components and has an extended production range of three axial heavy trucks, small buses and light trucks, 1.5 tonnes and 3 tonnes. It has close to one thousand employees, producing an average of four thousand trucks a year, which covers around 80% of the total Libyan market. It is currently undertaking the proceedings in order to get the ISO 9001 certification. The company is included in the new privatisation process going on in Libya despite actually being a partially privately owned company, the reason for being included in the privatisation process is the willingness of the authorities to increase the volume of private shares within the company. The company has to offer very competitive prices, a solid infrastructure and facilities, accessibility to the components and new products within Iveco and a free hand to develop and create new products. The country's industrial sector is a beacon of opportunity; Libya is located in the middle of Africa being the gateway to the continent and has succeeded in keeping a good relationship with the other African countries. Commercial and industrial discussions are taking place at the moment. On another perspective of the collaboration between Libya and its neighbouring countries, the company just sent one hundred vehicles into the Darfur region as a support to the Sudanese crisis.

"Foreign investors should make use of the current opportunities, they will find in Libya very honest and qualified people!"

TOURISM

OVERVIEW

Libya enjoys valuable natural and historical resources over a large geographical area. Long sandy beaches on the Mediterranean Sea, a rich history of past civilizations, well-preserved archaeological sights, sunny days all year around, deserts and mountains are all incentives of a profitable tourism industry.

As the country is located in an exclusive strategic geographical position from a tourism perspective, it is the perfect gateway to Africa from any European city in just a few hours flight, being located in the middle of North Africa and bordering the Mediterranean Sea with a coast line of about 2000 km length.

Due to its location, the country has witnessed many civilizations and cultures throughout its history, unique and well preserved historical sites to be visited are scattered around the country. Libya benefits also from a typical Mediterranean climate along the coast in the northern parts and desert climate in the south, average temperature range between 15 °C during winter and 35 °C during summer time. With 85% of the local population living in the major cities, most of the land is virgin. Furthermore, there is a potential workforce to be trained in the tourism field, as 61% of the Libyans are under 25 years old. There are absolutely no religious conflicts in Libya; the country is one of the most secure in the area. The official language being Arabic, with English, French and Italian being widely used by most of Libyans.

The Libyan authorities are currently encouraging individuals to take the initiative and form small and medium enterprises. Also the Government is working on modernizing the banking system and financial institutions, promoting foreign capital investment in Libya for the establishment of small, medium and large investment projects within the broad lines of the country's economic and social developmental strategy.

Market analysis has demonstrated that, whilst the traditional beach holiday will remain an important component of tourism worldwide, increasingly tourists are seeking a broader and, arguably, more interesting product involving one or more of:

– Special interest pursuits
– Activity based holidays and
– Trekking/adventure holidays

An evaluation of the tourism potential tourism within Libya indicated that the country would possess a significant appeal to a wide range of international tourists. The principal attractions may be summarized as following:

– Classical archeological sites.
– Desert landscapes and numerous prehistoric sites.
– Mountainous areas, particularly those of the Jebel Akhdar, Jebel Gharbi and the Acacus.
– Historic towns and cities that exhibit a wide architectural and cultural background.
– Beaches and sea
– Other tourism resources, including opportunities for scuba diving, religious tourism, World War II history, etc.

Opportunities can be found in the development of domestic tourism, particularly through the provision of family oriented beach resorts and supporting tourism and recreational facilities. With the increasingly high proportion of young people within the population, there is an urgent need to provide places of interest to visit and things to do.

Although holiday tourists are attracted to Libya at the present time, their number is limited and would not support a viably sustainable tourism sector. To achieve this goal, the development of the tourism sector is required, in particular:

– Investment in new accommodation.

– Significant improvement both in the development and presentation of tourist attractions, including those targeted at the domestic tourism market.

– The provision of ancillary supporting tourist facilities and services.

Tourist Accommodation

Libya possesses about 120 hotels or other forms of tourist accommodation offering nearly 9,000 rooms or their equivalent. The majority is located in the urban areas of Tripoli and Benghazi but few attain acceptable international standards.

In a survey prepared by WTO in 1998, visitor arrivals have been translated into room requirements as follows:

Regions	Number of Bedrooms				
	2003	2008	2013	2018	Total
Western	2770	2470	3180	5160	13580
Southern	500	560	680	1160	2900
Central	250	260	310	620	1440
Eastern	1100	1160	1510	2660	6430
Total	4620	4450	5680	9600	24350

Incremental room requirements

A range of investments in new accommodation will match more closely the specific market requirements of the different tourist segments.

Close to its cities people are always surprised to discover the impressive sandy beaches that are unknown by most of the people

MAIN COMPANIES

RECOMMENDED PARTNER

Afriqiyah Airways

Capt. Sabri S. Shadi
Chairman
Waha Building-Omar Almokhtar St.
P.O. Box 83428
Tripoli
Tel: + 218 - 21 - 4449734
Fax: + 218 - 21 - 4449128
E-mail: sshadi@afriqiyah.aero
Website: www.afriqiyah.aero.com

This airline was created after the lift of the UN sanctions on Libya as a response to the need to link both the countries belonging to the African Union and Libya with the European countries. Afriqiyah (Africa in Arabic) links its name to the date 9-9-99, which is the date of the creation of the African Union which happened during the African Summit held in Sirte, Libya (9th September 1999). The airline started its first operations with a first flight from Tripoli to Khartoum (Sudan) on the 1st December 2001. The main and final objective of the company is to make Libya a major player in Africa by making Tripoli a gateway to the continent. Currently, Afriqiyah is operating a fleet of three Airbus 320 with a capacity of 140 seats each and a staff of 275 employees approximately. In Europe, there are available flights to Paris (France), Brussels (Belgium), Geneva (Switzerland) and London (U.K.). In Africa, the company is flying to eight major African cities including Cotonou (Benin), Abidjan (Ivory Coast), Bamako (Mali), Khartoum (Sudan), Ouagadougou (Burkina Faso), Niamey (Niger), N'djamena (Chad), and Lomé (Togo). Starting from November 2004 the airline will also offer flights to Accra (Ghana) and Lagos (Nigeria); also the company will increase it fleet by one more A320 in 2005 and another one in 2006. Also, the company has initiated talks with other

European carriers in order to achieve commercial agreements that could implement the network of Afriqiyah allowing further connections to other European cities, the same could be happening with the U.S. soon. Libya is only at a couple of hours flying from Europe, it is a very short flight which, together with the country's climate, huge coastline and welcoming society, is a major advantage. On the business point of view, Libya has a big potential, not only in the oil sector but also in terms of infrastructure development as well as in the tourism sector. Libya is definitely a great destination.

"Libya is a great country, it deserves people coming to see and explore it, all the excitement is here and combining it with Afriqiyah will insure the best Libyan experience!"

quality equipment. Soon a bowling centre will also be available. The equipment for music was actually used and installed at the first time when the park hosted the Superstar event, a really famous singing contest in all the Middle East and across many countries in the world. The company is also planning to open a new hotel with 50 rooms close to the park in six months in order to implement more and more the diversification and quality of the park services together with a Spa fitness centre with thermal therapy. The main advantage of the Alsanawbar Park is the privacy of the facility and its location inside the city, which is perfect for this kind of activity, together with the quality of services and equipment. The park welcomes any visitors, nationals or foreign, as well as any tour operator willing to cooperate with them on any event that could take place inside the park and in the soon-to-be hotel facilities.

Alsanawbar Park Tourist & Family

Fowaad A. Mouny
General Manager
Head Office in Alnaser Jungle, Tripoli
Tel: + 218 - 21 - 3609859 / 3606797
Fax: + 218 - 21 - 3606765
E-mail: alsanawbar@excite.com

Alsanawbar Park Tourist & Family was created in 2004 as an answer from the Wedjan Company for Cater & Tourism Service for the need of these type of facilities due to the lack of leisure establishments in Libya. This is the first project of the company to be achieved but there is another one to be done soon by the sea. Inside the Alsanawbar Park, where up to three thousand people can be seated, we can find facilities for business congress or general events such as family celebrations, weddings and banquet lounges. The park also has restaurant and café facilities; different leisure activities like pool tables and open-air wide screens and finally a concert area with modern and high

Board for Development, Expansion & Promotion of Handicrafts

Mr. Abdoussalam Farag Shogman
Secretary of the Board
Gherian
Tel: + 218 - 21 - 4891799
Fax: + 218 - 21 - 3612492

Each city in Libya has its own kind of handicrafts, which can be found in every house anywhere around the country. This is the reason the board was established back in 1999 with the main objective being to preserve and develop these national handicrafts of Libya. The main daily duties of the board inside their installations of Gherian are the production of handicrafts and the training of men and women students in this field. Within the activities in training and production of handicrafts inside the complex owned by the board in Gherian, you can find traditional Libyan carpet weaving, pottery and copper handicraft. Handicraft is an important industry as it is related to many different resources, but it is not easy to provide figures about the sector due to the difficulty in estimating the domestic production of the handicrafts. Libyans produce most handicrafts at home and it is certain that there is large domestic production of traditional handicrafts, which are an important asset for the tourists coming to the country interested in traditional Libyan handicrafts. This is why it will be an important market in the near future. Every North African country has its own particular handicrafts. Training young students in the art of handicrafts is very important, as one of the main objectives of the board is to preserve the national handicrafts. There is a need to provide the required skills needed for their production. Colours and techniques are different from other kinds of handicrafts, so the board seeks to focus on the originality of the handicrafts and not on mass production.

"Since handicraft is an historical heritage for the Libyan people, it is our duty to present it to the world; we must present it in such a way that it should convey the message by itself."

Buraq Air

Mitiga International Airport. P.O. BOX 93149, Tripoli
Tel: + 218 - 21 - 3500821 / 3510016
Fax: + 218 - 21 - 3500949
Website: www.buraqair.com

Buraq Air is the first private airline in Libya, founded in November. The company headquarters are located at Mittiga International Airport in Tripoli, Libya. Buraq Air is proud of its well-trained staff and employees and recognizes their valuable contribution. The company's success is based on the sense of belonging, professionalism, commitment, hard work and the team spirit of its entire staff. The staff at Buraq Airlines provides reliable, convenient and consistent air transportation that meets or exceeds customer expectations. Buraq Air is customer-oriented, pro-

viding top quality services at economical cost. It implements an effective professional strategy to fulfil its obligation in providing reliable and efficient air services. The company is engaged in international, domestic, charter flights for both passengers and cargo. In addition, Buraq Air provides consultancy, training and aviation management for executive and private companies. Buraq Air has been in the market since November 15th 2001 and by the first quarter it has reached a load factor of 76%; by the year end it topped 83% with dispatch reliability of over 99% in its domestic network. The company, in a short time, has been able to provide its own ground handling for its fleet in both Benina International at Benghazi and Mittiga International Airports at Tripoli. In less than one year the company managed to increase its fleet from 2 B 727 aircraft to 10 aircraft including its owned 3 Boeing 727-200 Advanced equipped with modern equipment and interior furniture and one IL76 (cargo) with max payload of 50 tons.

Corinthia Bab Africa Hotel

Joseph Pisani
Director
Souk Al Thulatha Al Gadim, Trípoli
Tel. +218 21 335 1990
Fax: +218 21 335 1992
Website: www.corinthiahotels.com
Email: tripoli@corinthi.com

Ideally situated on the North Coast of Africa, bordering the Mediterranean sea to the North and the Grand Sahara to the South, and offering new and exciting investment opportunities, Libya became the natural choice to Corinthia' s Mediterranean roots. The Corinthia group designed and constructed the Corinthia Bab Africa Hotel, and this unique building is a distinctive landmark in Tripoli. It combines a balance between modern architecture and traditional Arabic design. All 299 guestrooms and suites are exquisitely furnished, encompassing contemporary design with a more traditional touch, and specially designed furniture. World-class standards and services offered at Corinthia Bab Africa Hotel aim at setting the standards in the region and are sure to impress the most demanding business and leisure travelers. The Hotel offers a wide range of services to efficiently accommodate the needs and demands of the modern day business travelers, such as private fax machines, direct dial telephone with multiple lines, voice mail, data port, and ISDN lines on request, among others. In the short span of one year, Corinthia Bab Africa Hotel has already hosted major international events and has conference and events facilities that include a number of exclusive halls and meeting rooms. With flexibility in mind, catering and conference professionals will ensure that event hosters experience effortless planning. IT and Business Support Services take Corporate Meeting standards to new levels of excellence. For the smaller, and perhaps more private meetings, the Executive Club Lounge offer effectiveness in style. For those who wish to relax after a long working day, nothing is better than a visit to the Athenaeum Spa, where one will recover in lavish comfort. Private treatment rooms, fitness training, massage, indoor swimming pool, sauna, Jacuzzi, and other amenities make this a mystical oasis for the perfect revitalizing experience.

"We are here to cater to all our clients needs and exceed their expectations."

Department of Antiquities

Mr. Ali E. A. Chadouri
Chairman
Tripoli Museum, P.O. Box 892, Tripoli
Tel: + 218 - 21 - 4440167 / 330298
Fax: + 218 - 21 - 3330292

The main objective of the Department of Antiquities is to study, protect and preserve archaeological heritage as well as restore the monumental buildings of the country. Another task of the Department of Antiquities is to make other people aware of the archaeological treasures and historical heritage of Libya. In order to accomplish this task the Department has tight relationships with European institutions and universities. Currently, there are a large number of scientific missions working on archaeological fields in the country. There are ten Italian missions studying civilisations, which lived in the past history of the country. There will be many others coming from France, Great Britain, United Sates, Poland, Germany and maybe Spain, which just sent a proposal to reach an agreement with a university there in order to study Libyan historical treasures. The relationship with the European institutions is old, there have been pieces transferred from Libya to many different European museums in order to show them abroad and more recently the association led by Saif Al Islam Gaddafi organised an exhibition that was shown in Italy, Great Britain, France and Germany. The Department of Antiquities is ready to do more of these kinds of projects; it is a member of many universal organisations including the UNESCO, which declared five Libyan cities as protected human cultural heritage.

"The facts included in this guide book are brief and simple; any reader of this text when coming personally will find more and more fascinating facts on the archaeological and cultural heritage of the country. Libya is a meeting point of different civilisations; it is reflected in the heritage through different sources. Any visitors that come to this country will see how it participated in the formation of modern civilisation. As for the present time, the Man Made River Project is the latest Libyan gift to the history of civilisation, aiming to overcome the dryness in the north, as it is explained in the third universal theory, the Green Book"

Main Sectors of the Economy: Tourism

Libyan Arab Airlines

Dr. Hosin A. Dabnon
Chairman
Ben Fernas Center, Tripoli Airport
P.O. Box 2555
Tripoli
Tel: + 218 - 21- 3614102 / 3614824
Fax: + 218 - 21 - 3614815
E-mail: h.dabnon@ln.aero
Website: www.ln.aero

Libyan Arab Airlines was established in 1964 and has become one of the most reliable air carriers within the region. It has extended its network to most of Mediterranean cities and, as such, the airlines has been rewarded the bronze medal at the Rome airport. From 1992-1999, Libyan Arab Airlines only had domestic destinations and was forbidden to fly outside of Libya. Now, LAA is back on track with ambitious plans to redevelop the airline. With 1 million passengers in 2004, the goal for 2005 is to double the number of passengers. LAA is now reorganising itself to meet the local and international demand; new destinations will be implemented in 2005 (Far East, Africa) and new planes will join the fleet. With the opening up of Libya, LAA wants to be at the top to serve all the people coming to Libya or travelling into the country, LAA's management has launched many studies and strategies to be a leader in the market. Several agreements have already been signed with Swiss and Austrian airlines in different fields. LAA is looking for more partners in different fields of activities. The staff is trained up to international standards, the pilots and engineers have a reputation for being excellent, and new training programs are on the way as well.

Libyan Travel & Tourism Company

Mr. Fituri M. Saled, General Manager
Tripoli
Tel: + 218 - 21 - 3503282 / 3504464
Fax: + 218 - 21 - 3503282
E-mail: info@libyatravels.com
Website: www.libyatravels.com

Libyan Travel & Tours Company was the 1st tour operator and travel agent created in Libya. The main company objective was to contribute to the promotion and development of local and international tourism. The company offers a whole range of services provided as an incoming and outgoing tour operator such as archaeological and desert tours; adventure and off track trips; leisure and special interest groups; conferences and seminars; booking and issuing of air tickets through their offices in Tripoli and Benghazi as well as booking of all categories of accommodations; youth travellers; business and third age tourism. It also offers airport services of meeting/assistance and transfers; luxury land transportation; and cruise vessel's services and excursions among others. These services include all land arrangements from visa invitation to transportation, hotel reservations, and qualified multilingual guides to escort the travellers in the classical, cultural, study and desert tours. The company offers custom made tours and claims to have a competitive advantage based in its quality and reliability of services as well as the fact that they keep their promises. The company is communicating its activities through the Internet, as well as the most famous international fairs such as Munich, Paris, Milan, London and Madrid. Last year the estimated amount of travellers that visited Libya through the company was 2000: 60% Italian, 30% French and the remaining from many nationalities such as Japanese and Spanish. The company has big expectations for the future and bases its strategy in developing more quality and price competitiveness. Currently, there is a project for 300 bungalows near Leptis Magna.

Libyan Youth & Student Travel Company

Mr. Ashraf S. Shah
General Director
Ali K. Zaidi St. Scouts Building, Tripoli
Tel: + 218 - 21 - 3335776 / 4443135
Fax: + 218 - 21 - 3344296
E-mail: ashah@cts.ly
Website: www.cts.ly

The Libyan Youth and Students Travel Company started its activities in 2004 as the first and unique Libyan-Italian joint venture in the national tourism industry, with 51% Libyan and 49% Italian ownership. The domestic owners are the Boy Scouts of Libya with 20%, a Libyan governmental tour operator with 21% and the General People's Committee for Youth with the last 10%. The Italian owners are made up of the Italian Central Tourist for Students & Youth (CTS) with 39% and Blue Panorama Airline with the remaining 10%.

The Libyan Youth and Student Travel Company provides all services for students and youth and it also promotes the development of student tourism inside the country and all over the world. In addition, it offers discounts in all kinds of travel tickets. An example is that the company has an agreement with Afriquiya airlines for a 35% discount and with Libyan Airlines for a 40% discount for ISIC card-holders. Furthermore, there are discounts in restaurants, hotels, malls, museums, theatres and all youth attractions. In only 8 months the company has achieved agreements with up to 150 companies in order to provide discounts and benefits for domestic and international students in the country. The CTS, ISIC cards are the only cards approved by UNSCO and recognised all over the world. More than 5 million students use these cards in over 90 countries. The company not only targets the students market, it is actually investing in Valtour, the biggest tourist village company in Libya and is working on the implementation of its offers and services for a wide range of travellers, including busi-ness or incentive travels.

> **"Everyone is invited; this country is going to be a really important destination from one day to the next."**

Civil Aviation Authority

Dr. Mohamed Shlebik
General Secretary of Civil Aviation Authority
Tel: + 218 - 21 - 3605318 / 3330256
Fax: + 218 - 21 - 3605322
E-mail: shlebik@lycaa.org
Website: www.lycaa.org

With approximately 3,400 employees the Civil Aviation Authority is the institution monitoring the regulations of the industry and establishing, upgrading and running 13 airports, of which 5 are international. The immediate aims are upgrading and expanding the infrastructure, especially communications, construction and refurbishment of air-ports and runaways for the many of the tourist sites that can be found in Libya. The projects department is commis-sioned to conduct and prepare engineering and technical studies to implement the Libyan Civil Aviation plans and projects in accordance with local and international specifi-cations, related to airports & airfields construction, estab-lishing ATM systems & their related utilities, planning, designing and developing the CAA utilities, services and facilities, supplying & provision and supervision of all oper-ational requirements, suggesting development plans and their predicted budgets, follow-up of regional and interna-tional organizations activities & participating in the associ-ated conferences, suggestions of its services encountered fees and Implementing its staff training and up-grading programmes. The department is functioning through four specialized sections: Airport engineering section, Electrical & Mechanical engineering section., Communication & Nav-Aids engineering section. Airports affairs supervision

THE BUSINESS OF PLEASURE

Located in the heart of Tripoli, just outside the historical Medina walls, the five-star Corinthia Bab Africa Hotel helps make business in Libya a pleasure. Elegantly-appointed luxurious accommodation, extensive business and leisure facilities, exquisite fine dining options and the commitment to our guests encapsulated by the Spirit of Corinthia make this stunning hotel the first choice for discerning visitors.

CORINTHIA
BAB AFRICA HOTEL
★ ★ ★ ★ ★
T R I P O L I

فندق كورنثيا باب أفريقيا
طرابلس

New Horizons, New Challenges

الخطوط الجويّة العربيّة الليبية

LIBYAN ARAB AIRLINES

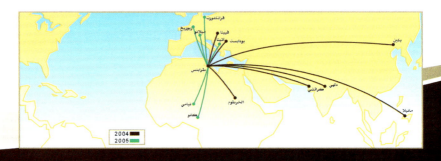

BEN FERNAS. CENTER TRIPOLI AIR PORT

Tel: +218 21 361 4102 - 361 4824 **Fax:** +218 21 361 4815 **P.O. BOX**: 2555   Sita: Tipzzln   Tripoli - G.S.P.L.

Section.

"Libya is the right place for investments, and the moment is right. The market is in need of know-how, and technology and the resources are here"

Rayah Travel & Tourism Co.

Mr. Nuri Mustafa Serrag
Chairman
P.O. Box 12051 Tripoli
Tel: + 218 - 21 - 3337508
Fax: + 218 - 21 - 3337508
E-mail: info@rtt-ly.com
Website: www.rtt-ly.com

The General Public Committee of Tajura initiated Rayah Travel and Tourism Company in 2004. The field of activities of the company is the possession, ownership and operating of tourism facilities such as hotels and tourism villages, recreation halls and beach resorts, restaurants, cafés and also providing catering services. As a travel agent itself, the company works on services such as undertaking exchange, insurance works and providing tourists special services. These services include selling, issuing travel ticketing and booking reservations for all types of transportation; organising tourism exhibitions, conferences and festivals inside and outside Libya, and operating tours for those coming and going including Hajj and Umra. The company also works within the sector itself on supplying equipment and materials related to the tourism activities. Within the sector the company deals with investment and ventures with national and foreign firms in tourism projects as well as organising tourism exhibitions, conferences and festivals inside and outside Libya. It also provides tourism promotion and propaganda services; buying and possessing of stocks and ventures in the establishment of tourism companies and investment and operating of all tourism transportation. Currently, the management is working on the expansion of the company by contacting international tour operators and welcomes any offer from them to collaborate in any tourism-related activity in Libya.

Tibesty Company for Hotels

Mohammed Ibrahim Tweir
Complex Manager
Khalig Sert St. P.O. Box 9200
Benghazi
Tel: + 218 - 61 - 9081071
Fax: + 218 - 61 - 9081079
E-mail: info@tibestyco.com
Website: www.tibestyco.com

Tibesty Hotels Operation and Management Company was establishment in 1989. The company is owned by the Libyan Social Security. The company operates and manages the Tibesty***** Tourist Complex in Benghazi, the Ouzu**** Hotel in Benghazi, the Garounis*** Resort in Benghazi, the Sirt***** Conference Complex in Sirt and the Al-Masrah Hotel in Tobourk. The company provides first class services in many national and international conferences and meetings. Among the type of services provided are room and board in its hotels, catering services for social institutions, catering and management for conferences and restaurants and supplying and delivering food. The company's goal is to improve and provide high quality services through development and training of its employees. In addition, the company aims to expand its operations to other parts of the country by managing hotels in cities such as Derna and Ras Hilal among others.

Wahat Fezzan Co. for Travel & Tourism services

Fnayet Al. Koni
Chairman
International fair, 225 Tareq street, P.O. Box 91863, Tripoli
Tel: + 218 - 21 - 3333366

Fax: + 218 - 21 - 3333366
Email: fezzanoases@hotmail.com

Wahat Fezzan Co. has had one main office located in Ubari City (southern Libya) and two branches in Tripoli and Ghadames since 1995. Some of the activities of the Company are the following. The management of towns, restaurants, hotels, amusement centres, cafes, camps, summer resorts and tourism centres. It also arranges tours for tourists and travellers and provides services to those tourists and travellers. It is involved in the sale and issuing of travel tickets in various means of transport and passport procedures and the invitation of groups of tourists from abroad introducing them to acquaint themselves with the archaeological places in Libya, and the organisation and arrangement for pilgrimage and minor pilgrimage. The company also organises exhibitions and festivals concerned with popular heritage and traditional handicrafts and participates in such activities inside or outside the country. It participates in the establishment of tourist projects with various related public and private sectors, and has the right to own estates and means of land and sea transportation facilities. The company has great experience about the desert and its population, the Tuareg; the staff knows and takes care of any detail in the great Sahara. Among those details, the company can provide tours with camel riding and 4x4 vehicles, car racing, motorcycling, mountain climbing and so on. Wahat Fezzan Co. also has a project to build a tourist club in Zuara with the collaboration of Italian tour operators and another project of realising a hot-water training centre located in Ghat. In the future, the management of the company plans to provide Libya with more and comfortable tourist facilities and to carry out a complete promotion of Libyan treasures all over the world.

"We are welcoming all the investments in Libya, particularly in tourism. Libya is a real chance for any tour operator who wants to handle the opportunity of a new tourist destination. Many silent treasures are waiting to be visited and promoted."

RECOMMENDED PARTNER

Winzrik – Group

Mr. Abdurrazag Gherwash
Chairman & Managing Director
37/32 7th November Street, P.O. Box 12794, Tripoli
Tel: + 218 - 21 - 3611123 / 25
Fax: + 218 - 21 - 3611126 / 3338364
Email: gherwash@winzrik.com
Website: www.winzrik.com

Winzrik Group was established in January 1991 with 138 experienced employees. Currently, with 380 employees, the company is considered one of biggest tour operators in Libya and is the first to work with tour operators of all nationalities with whom they have a very successful working relationship. Proof of this is seen in that they were in charge of organising the journey in Libya of the first Americans coming by ship to Tripoli. The company provides different services as group operator, travel agent, tourist village & hotels operator and tourist transportation among others. Winzrik also owns and runs the Winzrik hotel located in downtown Tripoli, with 48 rooms and 1 suite, including a restaurant, coffee shop and business centre. The Company also owns a motel in Ghadames located in the old city as well as the Tasili Hotel located in Ghat and

The tourist village in Janzour is one of the favorite places where Libyans go during summer

AFRIQIYAH AIRWAYS

Trust us to carry you away.

We provide the very best
check-in and inflight services,

www.afriqiyah.aero

Take advantage of our competitive fares,

We can fly you to 16 International
destinations in Africa and Europe,

Relax and enjoy the comfort
of our spacious Airbus A320,

**Fly Afriqiyah Airways and make all
your journeys a success.**

London
Brussels
Paris
Geneva

Tripoli

Khartoum

N'djamena

Bamako Niamey

Ouagadougou Lagos

Abidjan Lome
Accra Cotonou

the Al-Jazeera Tourist Village in Misurata. Winzrik also manages and owns camps and bungalows in other locations and is investing in many different lands across the country purchased strategically for development in response to the growing tourism potential of the country. The company is constantly striving to improve the overall quality of the tours by offering more efficient service and better facilities to all its clients. Winzrik is a member of Libya General Board of tourism, ASTA, DRV, WTO and Trade leaders' Club. The competitive advantage of the company is its quality: an example of this is that all of the drivers and experts have a wide expertise of the terrain and the guides are multilingual and professional in the field of tourism. Services provided are in continual improvement and they now also include issuing of entry visa at the boarder of their clients. Winzrik was awarded during the 27th international trophy for quality 1999 & the 24th international award for the best trade name in 1999, receiving trophies granted by the editorial office and trade leaders.

"Winzrik is your best partner in Libya."

The Bab El Jadid Hotel in Tripoli, in front of Al-Fateh Towers

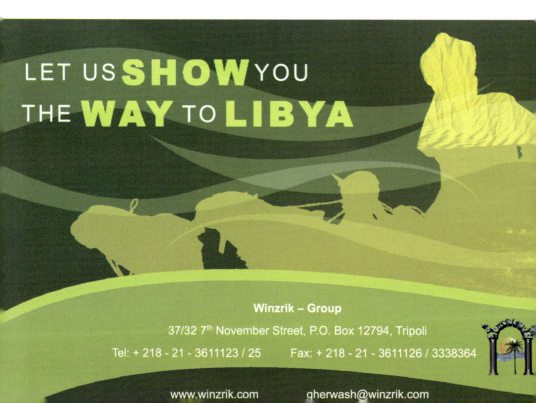

LIST OF SOME OF THE 60 MOST DYNAMIC COMPANIES AND ORGANISATIONS IN LIBYA

- Academy of Graduate Studies
- African Sail Shipping Co Ltd
- Afriqiyah Airways
- Akida
- Alcatel
- Alghad IT Services
- Alsanawbar Park
- Aman Company for Tires
- Arab Union Contracting Company
- Arabian Gulf Company (AGOCO)
- Ashoula Battery Co. Ltd.
- Azawia Oil Refining Company (ARC)
- Brega Oil Marketing Company (BOMC)
- Buraq Air
- Corinthia
- Electronics General Company
- Emgeg Consulting Inc.
- Engineering Industries Company
- ENI GAS
- General Tobacco Company
- Hydro
- Jowef Oil Technology
- Kanoun & Co.
- Libyan Arab Airlines
- Libyan Arab Domestic Investment Company
- Libyan Arab Investment Company
- Libyan foreign invesmtent board
- Libyan Insurance Company
- Libyan Travel & Tourism Co.
- Libyan Youth & Travel Company

- Maya Studios
- National Board of Handicrafts
- National Development & Building Materials Industry Company
- National Investment Company
- National Oil Fields & Terminals Catering Company
- National Oil Wells Drillng & Workover Company
- National Trailer Company
- Overseas Shipping Company Libya
- Petrolibya Oilfield Services
- Phoenicia Group
- Ras Lanuf Oil & Gas Processing Company
- Rayah Travel & Tourism Co.
- Repsol YPF (REMSA)
- Sahara Bank
- Schlumberger
- Sirte Oil Company
- The Libyan Businessmen Council
- Tibesti
- TNT
- Tripoli Int Fair
- Tripoli International Fair
- Trucks & Buses Co.
- Umma Bank
- United Insurance Company
- Veba Oil Company
- Waha Oil Company
- Wahat Fezzan Co.
- Wintershall
- Winzrik
- Zueitina

LEISURE

OVERVIEW

Libya enjoys a variety of tourist resources spread throughout the country. These include natural landmarks and sites such as coastal beaches, rocks and lakes overlooking the Mediterranean Sea, the Green and Western Mountains, as well as the many natural sights in Libya's desert areas. Human achievements in the field of construction, buildings, historical arts and cities, traditions, customs and folklore are considered of great significance due to their origins which date back to pre-historic times.

Plants and animals, although rare, are a part of a great number of tourist attractions in Libya. These include desert plants, turtles of multiple heads and a large variety of birds such as falcons and partridges. In general, the main tourist resources are classified as follows: desert, beaches and sea, old architecture sites, historical cities and mountains. Additional tourist activities include sea diving as well as health and religious areas and sites. The tourist areas in

Libya may be divided and/or classified according to their geographic location: Coastal Area Tourism, which includes beaches and the sea, old architectural cities, historical cities, Islamic landmarks etc, Desert Tourism, which includes a variety of natural desert sites, a huge treasury of pre-historic arts, agricultural and developmental structures in villages and settlements in oases areas, as well as the distinguished culture and folklore in oases and desert cities; and Mountain Tourism, which includes natural resources of beauty, distinguished culture, architecture, historical arts and unique civil constructions.

Beaches & Sea:

Libya overlooks the Mediterranean Sea with a coast line around 2,000 km long. This makes Libya's resort beaches among the longest of the Arab and African beaches along the Mediterranean Sea. Libya's beaches are characterized by variations in their natural appearances from sandy

Libya has over two thousand kilometres of coast, so take a swim and relax

beaches under the shadows of palm trees to rocky beaches overlooked from above by great hills, such as the one that can be enjoyed at the Green Mountain Sea Resort. The Libyan coastline, in general, is distinguished by its flatness with the exception of some parts where one can notice several topographical and morphological aspects such as the coastal coves and zigzags, rocky tongues and armlets, small islands, hills and rocky protruding formations, which taken together create a dramatic, natural coastal appearance. One of the most obvious of Libya's gulfs perhaps is the Gulf of Sirte. The most famous of sea heads is Ras-Gdear Head and the largest island is Farwa, near the western coast of the country.

Touaregs riding camels in the great desert.

There are many beaches spread along Libya's northern coast, the most significant of which are those found west and east of the cities of Zwara, Mellita and Telil. To the west of the old Phoenician city of Sabaratha, there are the Farwa Island beach, Tripoli, Khomes, Zletin and Misurata. These great beaches are distinguished with their soft white sand, tall palm trees and sandy hills covered with grass and forest trees, all of which form the backdrop. There are also several small beaches that provide tourists and beach lovers with even more great attractions, such as Psis and Nagaza, located between Garabouli and Khomes cities. Whilst in the north-eastern part of the country, namely in Benghazi, the second largest city in Libya, there is a great number of highly attractive beaches with soft, clean sand. Many of these beaches were developed and promoted to become attractive tourist resort beaches to absorb the great number of tourists, local beach lovers and holiday travellers.

In north-eastern Benghazi, from Touchier to Tolmitha, which is distinguished by its vast beaches, there are beaches which are wider and equaly attractive despite their small size. These beaches could become international destinations for tourists from all over the world. In addition to the aforementioned, there are several attractive beaches lying between Alhena and Ras-Alhamama near Sousa. East of Sousa, towards Derna city, there are splendid coastal

sights. The beaches located near the urban areas of Tripoli and Benghazi attract local inhabitants during the summer, where beach resorts and tourist villages are considered the most loved elements of the local tourism industry.

In general and without dispute, Libya's beaches provide a great variety of enjoyable opportunities, particularly the beautiful natural sites and the different water sports available, including diving and sea sports using vessels of all types.

Desert:

Natural desert makes up about 90% of Libya's total area and is an important source of tourism, as it is home to several attractions and landmarks of great importance. Such landmarks consist of different varieties of natural sites, agricultural and urban constructions in the villages, settlements and oases and desert lakes, in addition to huge, prehistoric arts and culturally important folklore art. This wide variety of tourist attractions provides many opportunities to pursue recreational, cultural and scientific activities, which can fulfil the wishes of the adventurous, athletic or mature tourist. There are activities for those who love to cross the desert through its many paths and passes, for those who look and long to discover every possible new thing, and for those in search of human culture, history and nature. Also, the natural beauty of Libya's deserts, their isolation, tranquillity and simplicity of life, has a unique quality which attracts photographers and many others seeking peace and privacy.

Aspects of Tourist Attractions of Libya's Deserts:

The desert areas in Libya have a mild to warm climate most of the year. The area hosts a large variety of natural geographies from mountains to massive colourful sand dunes to lush oases. There are the Akakous Mountains, the Elawinate Mountains, and the Melliet and Staft Passes.

Impressive sand dunes can be found in Ubari and Murzak, as well as great sand, sea and rocky terrains such as Alhamada Alhamra (the Great Red Hill). Some of the most popular oases are: Ghadames, Ghat, Wadi Alhyat Ashati Aljofra and Kufra. Besides the aforementioned natural landscapes, there are also other desert phenomena not found anywhere else, such as desert lakes in the Azalaf desert area, silent volcanic areas such as Wawo Enamous and rock formations containing multiple natural aspects such as pillars, rocky tables and others, all formed by erosion.

In general, oases are considered among the best tourist attractions of the desert as they usually exist in low altitude areas where water sources are near the surface. As a result, these oases tend to be lush with plants and trees, which usually cover quite large areas. Such trees are usually palm trees, surrounded by sand dunes or great lakes both of which provide special options for the

Many ancient structures can be found deep in the desert

The Gabar-Aoun desert lake, an emerald in the middle of the desert

where its fringes are decorated with an enclosure of palm trees, bamboo trees and Tamarisk trees, some of which can be seen from the sand hills.

These sand dunes and hills offer adventure to drivers due to the great pleasure of driving over sandy hills and areas with 4x4 vehicles. From the top of the sand dunes and hills one can watch distant oases and enjoy other beautiful sights.

tourists. These oases are also rich in cultural heritage, with their old cities and villages. Oases are known to be green areas in the heart of the lonely desert, which throughout the ages have provided water and food to their inhabitants, enabling them to carry out social and economic activities. As a result, civilizations and cultures in the heart of deserts were formed and existed for centuries, the most significant being at the Ghadames, Ghat, and Murzak Oases.

The breathtaking sand dunes found in Libya's deserts, which cover a large area, form some of the most distinguished and beautiful landmarks. Tourists and travellers will no doubt be surprised by the simplicity and organization of these varied formations as opposed to merely being scattered piles of sand. There are the "crescent form" sand dunes, "star-like" dunes, "domed" dunes, "net-like" dunes and "longitudinal" dunes (sword-like). These deserts and sand dunes, in addition to their great natural formations and beauty, offer the tourist places and opportunities to exercise and perform his/her desired type of sport such as sand skiing/boarding and a variety of sail sports among others. Other benefits of sand dunes are that tourists can use them as hot, sandy baths such as the ones found in the Ghadames Oasis, a great tourist attraction.

There are also the lakes in the sandy hills of Dawada near the Ubari Sand Sea in Zalaf Sand Hills (Ramlat). These are also great tourist attractions within the Libyan desert,

The most important desert lakes are Gabar-Aoun Lake, Mandara Lake, Umm Alma (Water Mother) Lake, which is one of eleven lakes located in Zalaf Lake Area, and the most attractive for tourists and visitors. There is also Bzima Lake surrounded by Rubyan Sand Hill near Kufra Oasis, which is yet another great tourist destination.

Mountains and desert hills in southern Libya add yet another ingredient of desert beauty. The most striking of these are in the Akakus Mountain area in the southwestern part of the country near the Libyan – Algerian border. These mountains extend north to Alwinat and south to southern Ghat, parallel to Tinzuft Wadi. This area enjoys splendid natural beauty due to the many caves, which contain aspects of pre-historic artwork. The Akakus Mountains are also distinguished by their multi-coloured rocks of steep margins which give them a very majestic look. In general, the Akakus Mountain Range, because of its attractive variation in rock colour, size and formation due to erosion and volcanic rocks, is the most impressive and most beautiful natural landmark in Libya.

The black rock (Harouj) is another aspect of the attractive, natural geography of Libya, due to the great variation in its volcanic rocks and crimpy terrain. In fact, Harouj forms the largest area covered with dormant volcanoes in all of Africa. It is located in the middle of Libya and can be reached from many different directions; from Zala oasis in the north, Afokha oasis in the west and Tamessa in the southwest. It seems that the black Harouj was a pasture for both wild and domesticated animals, which used to live by the lakes and water paths during pre-historic times. Many drawings were discovered on the rocks representing wild animals such as lions, elephants and giraffes along with many sites for making sharp stone tools, which can be traced back to the modern and old stone ages at Algadari Wadi, Ashadid a Wadi north of Alharouj, north of the seventh Gour. Wawo Enamous Volcano, along with its surrounding lakes has been described by writers and travel journalists as one of the most beautiful natural sites anywhere in the world. Wawo Enamous is located in an isolated area 100 km south of the black Harouj in the middle of Tebisti Sarir plain. Wawo Enamous Volcano is surrounded by about 10 splendid lakes some of which have pure water springs surrounded by palm trees, long bamboo canes and Tamarisk trees. Some of these lakes are red in colour due to the many types of Crustacea growing in them.

There are several types of bird species, foxes, wolves and reptiles that can be found in this desert region. In addition to these desert mountain hills, there are mountain areas such as Alwinat, which are eastern mountains nearby the Libyan – Egyptian border, which especially attract visitors and tourists who wish to watch the free falcon and the splendid mountain terrains, as well as other less important hill sites.

Libya's deserts also offer a wealth of dry rivers (wadis) which represent a clear comparison with the dryness of the desert region. Such areas are usually rich in underground water in the desert thus allowing them to become the most settled and inhabited places throughout the ages. Apart from their natural sites and vegetation, which includes medicinal plants, palm tree, grass and desert trees attracting researchers and nature lovers, there are many caravan paths which come in the form of wadis and sandy paths. These caravan paths cross the desert and usually lead to old cities, the most famous of which are Wadi Ashatti and Wadi Alhayat in south western Libya. Both of these places are home to Germent tribes, which established and built a desert civilization that can still be seen today. All of the above, in addition to Wadi Attaba, Wadi Barjuj and the Akakous Mountain's wadis preserve desert pillars, rocky tables and false castles are some of the most beautiful natural sites in Libya's deserts. These areas are full of plant and animal life, distinguished by their ability to withstand natural conditions such as dryness and intense heat most of the year. The plantation cover in the Libya's deserts is generally characterized by variations of Tamsrisk trees, Acaci trees, as well as a great variety of rare medicinal plants and grasses.

With respect to animals, there are deer and wadan (deer-like animals) as well as a great number of birds and insects all of which represent an attraction for tourists, especially those who love nature and enjoy eco-travel and research.

Libyan art can be found even in the most curious places, here for example as a drawing in the desert sand

174

Warm and peaceful people, the hospitality is a tradition for the Libyans

Historical and Cultural Tourist Attractions of Libyan Desert:

Libya's desert area is famous for its size as it represents a large part of the great desert. Its strategic location as the crossroads of old caravans between equatorial areas in the south and the mild areas in the north, as well as its proximity to the Nile River, Eastern Africa and Mediterranean civilizations, has had a significant cultural and humanistic effect on the area. The human settlements in these areas go back to pre-historic times, evident by inscriptions and drawings found on rocks and in caves in the mountains and valleys, as well as the man made handicraft tools used to fight wild animals and in hunting. The remains of old cities and villages also provide evidence that several great civilizations once existed all over this now rather silent desert area. Such civilizations included the Germent and Roman civilizations as well as the Arab Islamic civilization which now extends over vast areas of Libya's deserts. All these historical arts, archaeological sites, oases and similar artificial and natural sites represent a great and valuable treasury and an economic source of income in southern Libya in the form of tourism. Such treasures including the rock formations in the Akakous Mountains and Ghadames city have been declared as protected historical sites by UNESCO.

The Green Mountain:

The Green Mountains, Gebel Al-Akhdar, and the Western Mountain, Gebel Al-Gharbi, are considered the main mountain ranges in both the eastern and western parts of the country respectively. They are situated along the coast of the Mediterranean Sea or relatively near to it. Both areas are considered particularly high with a maximum altitude of about 1000m. Therefore, they are higher than the neighbouring coastal plain lands or the coast area itself and they also contain narrow and declined valleys. The rainfall rate in the mountain areas is considered relatively high; therefore plants and vegetables grow there more than in other areas of Libya. The base of the mountain and the foothills provide beautiful scenery, especially in spring when the temperatures in those two mountain areas are cooler than the surrounding plains in summer, but they are much colder in winter.

The Green Mountain, Gebel Al-Akhdar, area is characterized by its beautiful natural scenery.

Its natural environment is rich with plants and vegetation that attract many tourists. It also has geomorphologic formations such as caves and natural shelters in the mountain. One of the most famous of these caves is Ahwa Ftaih, where some collections of artefacts belonging to pre-historic man, who once lived in this area, have been found. Moreover, the natural coastal scenery offers excellent tourist attractions that provide an area to be enjoyed and explored. In addition, there are famous archaeological sites that have great importance in the area with respect to tourism.

The tourist capabilities in the Western Mountain area can be characterized by the local culture, the architecture and unique construction styles which were developed within the isolated mountain framework. Besides the magnificent mountain towns and villages in the area, the prominent and important tourist places are the grain stores in

Prehistoric art drawn in the desert rocks

Nallot, Kabaow, Qasr El-Haj and the ancient villages in Timojet, Jado, Farsata, Yefren, Qalaa Castle and the prominent settlements at the summit of the hills of the towns Tormisa and Jado. These towns lie in the middle of the mountain where a small museum is also located which contains impressive samples of artefacts and archaeological finds that show the history and heritage of the whole area.

The Cultural Heritage:

The Libyan people are very proud of their culture, which they consider to be among the very richest and most diverse in the world. Apart from the historical aspects such as the construction, architectural styles and forms and museums, there are many other cultural elements that are of special interest to tourists. Such elements are the followings:

- Handcrafts and handmade articles.
- Native cuisine and cooking.
- Arts and music, including paintings and carvings.
- Festivals and special occasions.
- Folklore, singing and dancing.

From a historical point of view, Libya was the source of many crafts such as the traditional jewels and the famous Ghadames slippers, as well as handicraft works that have attracted tourists over the years. The most important traditional hand made products that are still made in the country are the ceramic pottery pieces that are made in Gharyan. Other important handmade products such as plates, baskets and hand fans are made from the palm leaves and branches, which are famous in the southern oases. Leather products such as handbags, belts and saddles in addition to the steel products such as swords and daggers, are famous in the Ghat area. Different kinds of Libyan, Arab and Mediterranean cuisine and foods are available in Libya. Many tourists are interested in the method of cooking such specialty dishes.

Furthermore, the quality of the fruits, vegetables and meats that are available in Libya is considered high, as well as the basic foodstuffs, such as bread which is also considered excellent and very diverse.

Modern art is considered a successful sector which is now being developed within the tourist market. Many art galleries were established over the last few years where high-class, artistic works have been displayed, which attract tourists and visitors. Traditional music is also available to a large extent in the country and can be heard in many cultural festivals around the country.

Festivals and special occasions take place regularly in many places in Libya. The most prominent tourist festivals are those held yearly are in the cities situated in the oases such as Ghadames, Ghat, Hone, as well as in the Gebeal Akhdar and in the Western Mountains and in Tripoli city. The festivals that take place in the ancient cities in the oases are considered pleasant and fun. They attract tourists because they offer them a great opportunity to see traditional and folklore dancing, singing and music, in addition to the presentation of the local customs and traditions in an original and historical atmosphere.

Old Historical Arts:

Archaeologists have excavated a variety of objects and sites left by ancient civilizations which give evidence of the successive and rotating periods suffered by the desert area in pre-historic times. They have found inscriptions and drawings in hundreds of caves and mountain area wadis, especially in the Akakous Mountains and Metkhandush Wadi. These inscriptions and drawings left by our ancestors, while representing a small part of man's history, stand as a witness to a period in which the area was once rich with plants, grass and even different animal and bird species.

These old and beautiful works of art are highly valued and regarded as one of the best tourist attractions of the area. Akakous Wadis and Akakous Mountains have wide caves which were formed due to erosion and are regarded as the most important mountain area. These areas are full of and rich with pre-historic rock arts, which the artists had chosen as their fine arts centre, and which now represent the heart of pre-historic art in Libya and possibly on the southern shores of Mediterranean and Europe. This area has also been declared by UNESCO as an international heritage site of man's civilization. These pre-historic rock art and archaeological sites are regarded as a highly attractive aspect for different tourist groups who have different desires and intentions.

While such art with its aesthetic appeal and artistic level attracts tourists in general, we find that its date determina-

A cat drawn in the times of the ancient civilizations

The camel race is part of the festivities of the Ghaddames festival, where all the tribes from the desert meet once a year

tion, interpretation, analysis and whatever else might be acquired or obtained as information, is an additional attractive feature. This information can shed light on the environment, the economic base of those who carried it out and implemented it, the discovery of their origin and race, and knowing what living means they used such as cloths and war tools. The art can also tell us about their life style, living quarters, social environment and activities, as well as their faith trends if possible. All of these can attract tourists, archaeologists, geologists and those who are interested in old desert cultures.

The Ancient Ruins:

The Libyan coastline hosts the most beautiful ancient ruins

in the world, which are predominantly concentrated in both the eastern region and the western region of this coastline.

In the western region, the history of the ruins date back to the Phoenician era when the cities of Sabratha, Oea and Leptis Magna were founded. Those ancient cities became the most prosperous and beautiful cities at that time and left fascinating ancient ruins and heritage,

In the eastern area, the ancient ruins include the famous five Greek cities Shahat (Cyrene), the port of Sousa at (Apollonia), Talmitha (Ptolemais), Tukrah (Teuchira) and Morj (Barca). The beauty of those ancient cities is enriched by their location in the Gebel Akhdar area, which has an attractive natural landscape. There are also other ancient ruins that are interesting to tourists such as the unique collection of beautiful Byzantine mosaics that lie in the museum of Qasr Libya, and are considered the main international tourist attraction, as well as the Libyan Temple located in the Saltana area, which was built before the Islamic era. There are other isolated Roman ruins located to the south of Tripoli city in the area of the Western Mountain (Jabel Al-Gharbi) area, and others near Yefrin town, Ghadames city and in the Germa area, deep inside the desert.

The most important Roman ruins, the fortified Roman farms, graves and the big tombs, can be found 200 km southeast of Tripoli in the Girza area and are particularly interesting. In light of international standards for ancient cities located along the Libya coastline, they were declared an international human heritage site (i.e. Shahat, Lobda

The well preserved theatre of Sabratha will transport you to the past

and Sabratha).

Religious Tourism:

Religion has an important role in the daily life in Libya. It is worth mentioning that ancient ruins left by the three major religions, Judaism, Christianity and Islam that prospered during different historical stages, are represented by the architectural, religious and historical buildings and art left in the country.

The Islamic sites in Libya acquired their religious importance from their ancient history. Whereas the history of most Islamic buildings and ruins, that have an architectural importance in the first place, refer to the first stages of Islam's spread through North Africa.

Such buildings are represented by the historical mosques in old Tripoli, by the small mosques that were built with small rocks in the western mountains, and the ancient mosques existing in the desert oases. Such ancient monuments generally attract the interest of international tourists.

At the present time, Islam is considered the religion adopted by all inhabitants of Libya. All graves and tombs of the companions of the prophet are usually visited by the people in the towns of Derna, Zwaila and Ojla, as well as the graves of famous saints and religious clerics in the cities such as Zlaiten or Zwara.

Christianity spread in Libya during the Byzantine era. It is believed that Saint Marcus is responsible for the spread of Christianity in Libya. The ancient monuments of Saint Marcus' first church in Libya are located in the mountain area behind Derna city. It is expected that when the current excavation works are complete, it will reveal the full range of those sites to tourists and those churches will be one of the most prominent tourist attractions.

The historical period when Christianity dominated in Libya left a heritage of churches whose sizes range between big

Together with Sabratha, Leptis Magna was one of the most prosperous and beautiful cities the Roman Empire ever had

splendid churches, such as those found in Sousa (Apollonia), Shahht (Cyrene) and Leptis Magna, to small churches that are under ground or in the low land of the Western Mountain area.

Jewish antiques are relatively limited. However, the most important ancient monuments are the Jewish Temple in old Tripoli and the small Jewish temples in the Western Mountain area which were built in the local style. Those buildings in general are considered of significant architectural importance.

One of the most famous cities with important sites for the Islam and Christianity is Darna. It is located to the east of the Green Mountains (Gebel Akhdar), in the middle of a narrow level plain between the coast of the Mediterranean Sea and the base of the Green Mountains. The most prominent Islamic site existing therein is the Great Mosque, which was built during the Ottoman Turkish era.

In addition, there are the tombs of the companions of the prophet which are the focus of religious visits. There is also the church of Saint Markus in the mountain region in the upper part of the city, as well as the ancient mosques inside the caves.

The small market, public yards, and the mountain scenes are also considered tourist sites having importance with respect to the area visitors.

THE WESTERN REGION

Tripoli

Known as Tarabulus in Arabic and Oea in ancient times, Tripoli is the capital of Libya. Once known as the 'White Bride of the Mediterranean', Tripoli has lost much of its pristine allure, though its many historic mosques and lively medina retain a good deal of character. The Turkish and Italian colonial periods have also left a distinctive mark on the city's architecture.

Easily the most dominant feature of Tripoli is the Red Castle, on Green Square, which sits on the northern promontory overlooking what used to be the sea but is now a motorway and 500m of reclaimed land now separating the two. The massive structure is comprised of a labyrinth of courtyards, alleyways and houses built up over the centuries with a total area of around 13,000 square meters. Facing the castle across from Souk al-Mushir, the Ahmed Pasha Karamanli Mosque, built in 1738, is the city's largest. The Jamahiriya Museum, built in consultation with UNESCO, is set in Tripoli Castle and contains one of the finest collections in the Mediterranean region.

The medina, a maze of narrow alleys and covered markets, is the best shopping in the city and is the heart of Tripoli. As only a handful of tourists visit Libya, the souq has an authentic air and the goods on display cater to local tastes. The old walled city also contains all of Tripoli's historic mosques, khans (inns), hammams and houses.

Leptis Magna

In Roman times the Tripolitania province, called Tripolis, had three cities: Leptis Magna, Oea (now modern Tripoli) and Sabratha. Leptis Magna was the second Roman port in

The desert is a beacon of peace and silence, the perfect place to pray for many people

Africa and is considered the best Roman site in the Mediterranean due to its spectacular architecture and its massive scale.

The city was originally a Phoenician port, settled during the first millennium BC. Slaves, gold, ivory and precious metals brought it great wealth, supplemented by the rich agricultural land surrounding it. Roman legions ousted the Carthaginians after which the city flourished until the Vandals came in 455. The city flourished in the early Roman Empire, particularly under the rule of the Severan dynasty, a family of Libyan origins that governed for almost half a century. The Arab invasions swept away the last traces of Roman life, and in the 11th century Leptis Magna was finally abandoned to the sand dunes.

Today its ruins are wonderfully well preserved and restoration work continues. The Severan Arch, in the main entrance of the city, was erected in honour of Emperor Septimus Severus' visit to his hometown in 203 AD. During his rule many imposing structures were built. Near to the entrance, the marble and granite panelled Hadrianic Baths,

After visiting the medina, a walk in front of the sea during the Sunset is the perfect combination

Sabratha was a prosperous town during the 3rd century AD. It was well known for its ivory trade which was brought from central Africa via Ghadames and Fezzan. In 455 AD Sabratha was occupied by the Vandals who pillaged it and left it in decay. Justinian, the Byzantine Emperor, occupied it in 633 AD, and fortified it. He built a magnificent church, the mosaic floor of which is now incorporated into the town museum. New houses were built and streets were constructed over the debris of the old town. The Byzantine ruled Sabratha until the Arab conquest, after which time it was then used as a military post after their triumph over the Byzantine armies. The commercial activities of Sabratha were subsequently transferred to Tripoli.

The seaside setting, the honey-coloured stone, and the quality of the carvings make Sabratha one of the most magical ruins in North Africa. The theatre is the main attraction. It has been carefully reconstructed and it is even used currently as an arena for theatre and concerts. Other archaeological monuments include several public baths, temples, fountains, and mosaics. The museum contains an extensive collection of small objects of everyday use founded in the houses such as pottery, glass and bronze utensils.

the largest outside Rome, can be found. Also to be found are the partially covered nymphaeum, a shrine dedicated to the worship of nymphs; a pair of massive forums, similar in design and grandiosity to the imperial forum in Rome; the extraordinarily detailed basilica and theatre; and 700m to the west along the seashore the circus and amphitheatre, where chariot races and similar spectacles were held for the locals' amusement. There is an excellent, large museum, next to the main entrance of the site, containing pieces that range from Libya's prehistory to gifts given to Colonel Gaddafi.

Sabratha

The city of Sabratha lies on the coast 70 km west of Tripoli and was considered one of the main commercial harbours founded by the Phoenicians along the coast of the Mediterranean Sea in the 6th century BC. The town of Sabratha has grown up in between the ruins, adding a special charm to the place. It flourished under the Romans, when many new buildings were erected, from the 1st century BC to the 4th century AD.

Ghadames

Ghadames is located 683km southwest of Tripoli, close to the border of Algeria and Tunisia. It is one of the most famous oases in the Libyan Sahara and it earned the sobriquet 'Pearl of the Desert' back in the 1950s. Ghadames is also renowned for its characteristic architecture and its inhabitants, the Touaregs, are one of the most colourful people in the Jamahiriya. They ride on their very fast dromedaries called "Mehari" which can run as fast as 50 km per hour and are known as the guardians of the Sahara.

LEISURE

Ghadames was once one of the most important trading towns of the northern Sahara. Historians record that it was inhabited 4,000 years ago. Excavators have found Greek carvings in a region to the northeast of the city, as well as the mixture of Roman and Garamantes art and architecture. In the 3rd century BC, there was also a castle built for Roman soldiers. In the 7th century AD Omar Ibn-al-As sent an Arab Muslim battalion to Ghadames in order to make it a foothold from which the Arab Muslim armies could later spread throughout North Africa.

The UNESCO World Heritage listed the old city as the largest and most well preserved site in Libya. It is a labyrinth, lit only by occasional overhead skylights and open squares - a style unique to this part of the Sahara. Near the western entrance of the old town, the House Museum is an old merchant's house with its original furnishings and decorations still intact. The Square of the Mulberry is the old slave market - many locals are descendants of former slaves- near the D'jmaa al-Kabir mosque, whose minaret can be climbed for an excellent view over the town.

The Arch and the streets of Leptis Magna, few Roman ruins are so well preserved in the world nowadays

At the end of the last day of the Ghadames Festival, everyone meets in the dunes to watch the sunset

WHERE TO STAY IN THE WESTERN REGION:

RECOMMENDED

Corinthia Bab Africa Hotel

Souk Al Thulatha Al Gadim
Tripoli
Tel. + 218 - 21 - 3351990
Fax: + 218 - 21 - 3351992
E-mail: tripoli@corinthia.com
Website: www.corinthiahotels.com

Safwa Hotel

Baladia St. P.O. Box 5133
Tripoli
Tel + 218 - 21 - 3334422 / 4443257
Fax + 218 - 21 - 3332019 / 4449062
Website: www.safwahotel.com

Winzrik Hotel

P.O. Box 12794
Tripoli
Tel: + 218 - 21 - 3344407
Fax + 218 - 21 - 3342871

The Tourist Village Janzour

Janzour
Tripoli
Tel: + 218 - 21 - 48904430 / 35
Fax: + 218 - 21 - 4890521

Both comfortable and convenient, the tourist village Janzour is located in the middle of a magnificent park near a beautiful sandy beach on the Mediterranean Sea Complex of hotels and apartments. All rooms are equipped with air-conditioning, bathroom, telephone, colour TV with satellite, mini-fridge and automatic alarm clock. Specialised restaurants offering international cuisine, international coffees and a coffee shop will also help meet your requirements. It also hosts a swimming-pool, fitness centre, different amusement games and a conference centre/showroom for hosting seminars, conferences and banquets.

El – Khaber Hotel

Tripoli
Tel: + 218 - 21 - 4443879 / 4445940
Fax: + 218 - 21 - 4445959 / 3334608

El – Khaber Hotel borders the gardens and faces the cornice of Tripoli. Its 334 rooms/suites are all air-conditioned and equipped with private bathroom, telephone, colour T.V. with satellite, mini-fridge and automatic alarm clock. There is a specialized restaurant offering international cuisine and coffee shop. There is also a swimming pool and fitness centre; we are proud of our 400 seat conference and banquet hall, business centre, bank and parking.

Bab Elbahar Hotel in Tripoli

Bab Elbahar Hotel

Tripoli
Tel: + 218 - 21 - 3350676 / 3350410
Fax: + 218 - 21 - 3350711

Bab Elbahar Hotel is located in the magnificent park near the beautiful beach of Tripoli. Its 403 rooms/suites are all air-conditioned and equipped with private bathroom, telephone, colour T.V. with satellite, and clock. A specialized restaurant offering international cuisine and coffee shop will meet your requirements. There is also a swimming-pool and an open air cafeteria. We are proud of our 400 seat conference centre for hosting seminars, centre, bank and parking.

El Wahat Hotel

Tripoli
Tel: + 218 - 21 - 3334061 / 66
Fax: + 218 - 21 - 4445055

El Wahat Hotel is located in the centre of Tripoli City not far from Tripoli's international fair, shops, offices and banks. Its 304 rooms/suites are all air-conditioned and equipped with satellite, mini-fridge and automatic alarm clock. There is also a specialized restaurant offering international cuisine and a coffee shop.

Bab El Jaded Hotel

Tripoli
Tel: + 218 - 21 - 3350571 / 74
Fax: + 218 - 21 - 3350670

Bab El Jaded Hotel borders the gardens and faces the Mediterranean coast. Its 347 rooms/suites are all air-conditioned and equipped with satellite, mini-frig and automatic alarm clock. A restaurant offering international cuisine and coffee shop will also help meet your requirements.

Bab El Madena Hotel

Tripoli
Tel: + 218 - 21 - 3350650
Fax: + 218 - 21 - 3350675

Bab El Madena Hotel borders the gardens and faces the Mediterranean Sea. Its 204 rooms/suites air-conditioned and equipped with private bathroom telephone, colour T.V. with satellite, mini-fridge and automatic alarm clock. A restaurant offering international cuisine and coffee shop will also help meet your requirements. There is also a swimming pool.

Zaitouna Hotel

Bani-Walid
Tel: + 218 0322-60040-44
Fax: + 218 3614860

Zaitouna Hotel located in the centre of Bani-walid City not far from the central banks, boutiques, and shops. Its 34 rooms and suites are all air-conditioned and equipped with private bathroom, colour T.V. with satellite, mini-fridge and automatic alarm clock. There is a restaurant with international cuisine and a coffee shop.

Zaitouna Hotel in Bani-walid

WHERE TO EAT IN TRIPOLI:

Africa Restaurant

Ahmed Al Sharef / Al Dahra
Tripoli
Tel: + 218 - 21 - 4447000

Al Athari Restaurant

Old city Goss Markous
Tripoli
Tel: + 218 - 21 - 4447001

Al Gambare Restaurant

Al Baladia Street, at the back of the Grand Hotel
Tripoli
Tel: + 218 - 21 - 3341287 / 3333637
Fax: + 218 - 21 - 3341287
E-mail: algambare@hotmail.com

Al Saraie Restaurants

Alsaha Alkhadra (Green Square)
Tripoli
Tel: + 218 - 21 - 3334433 / 3334434

Al Badawi Restaurant

Al Baladia
Tripoli
Tel: + 218 - 21 - 3339995

Al Masabeeh Restaurant

Al Saidi Street
Tripoli
Tel: + 218 - 21 - 3337815

Al Nakhla Restaurant

Al Baladia Street
Tripoli
Tel: + 218 - 21 - 4445173

View of the millenary city of Ghadames, in the middle of palm trees

Al Safeer Restaurant

Al Baladia Street
Tripoli
Tel: + 218 - 21 - 4447064

Al Sayaad Restaurant

Al Shatti Complex
Tripoli
Tel: + 218 - 21 - 4774111 / 4774552

Al Sheraa for Fast Food

Hai Al Andalus, near Qaat Al Shaab
Tripoli
Tel: + 218 - 21 - 4773873

Al Sheraa Restaurant

Hai Al Andalus, near Qaat Al Shaab
Tripoli
Tel: + 218 - 21 - 4775123

Dentice Restaurant

Garagarish, Hai Al Andalous
Tripoli
Tel: + 218 - 21 - 4773986
E-mail: info@dentice-fish.com

Galexy Restaurant

September First
Tripoli
Tel: + 281 - 21 - 4448764

La Valette

Corinthia Bab Africa
Tripoli
Tel + 218 - 21 - 3351990 / 3351992
E-mail: lgatt@corinthia.com

Tripoli International Exhibition Restaurant

Omar Al Mokhtar Street, Tripoli International Exhibition
Tripoli
Tel: + 218 - 21 - 4441418

THE EASTERN REGION

Benghazi

Situated on the eastern edge of the Gulf of Sirte, Benghazi is the second largest city in Libya and is famed in military history as a supply base during World War II. It has been fought over and rebuilt so often that little remains of its Greek and Roman origins. After the Arab conquest, it was bypassed by inland routes further south, but it was revived as a Turkish fortress and then as a centre of operations for the Italian invasion, and this is reflected in much of the Italian style architecture. The city today displays little of its ancient heritage, as it was largely pummelled into ruin during WWII and rebuilt only after oil money began pouring in following the revolution. What it lacks in historical charm, however, it makes up for in location. Benghazi makes a great base for exploring the lush Green Mountain area and the numerous Roman ruins along the coast, and there are good beaches nearby.

Cyrene

Second in importance only to Leptis Magna, Cyrene ranks as the best preserved of the Greek cities of Cyrenaica and is the largest of the five cities of the region. Apart from the spectacular Greek ruins, its location high on a bluff overlooking the sea is stunning.

It was the first Greek colony in North Africa and it was founded in 631 BC by Greeks immigrants from the island of Santorini (Thera). The city gradually developed into a Kingdom noted for its love of science and philosophy. Famous among the philosophers of Cyrene were Callimachus, Carneader and Aristippus. The Kingdom of Cyrene reached its grandeur around 400 BC. Pindar, the ancient Greek poet, described it in one of his odes as "the city built on a gold crown".

The city covers a huge area and is still only partly excavated. The site comprises five main areas: the Temple of Zeus, the Roman city centre, the Greek agora and Roman forum, the Acropolis and the Fountain and Sanctuary of Apollo. Mosaics can still be discovered underfoot, and priceless statues often lie covered with creepers.

Apollonia

The city of Sousa was known as Apollonia during the Greek-Roman times and was the port nearest to Cyrene. It is located on the Mediterranean coast, 20 km northeast of Cyrene, and is connected with the city by a road excavated in the rocky ground of the area, and improved during Roman times.

Throughout history, Sousa was influenced by all the changes that also affected its neighbours. For more than a thousand years, Sousa was a prosperous port, known for the exportation of sylphium, a plant that played a great economical role during the first centuries of the Greek settlement, around 6th Century BC. Its importance in that period exceeded that of Cyrene and Ptolomais

There are Greek, Roman and Byzantine ruins in Apollonia. The Greek theatre, the Roman baths, and the Byzantine palace, which served as a government building in the 6th century AD, are a few examples. Other ruins include the burial sites, the military forts, the theatre- situated in a particularly picturesque location by the sea- and the eastern, central and western basilicas.

WHERE TO STAY

Ghouz Atteek Hotel
Misurata
(The hotel is currently changing its telephone lines)

Ghouz Atteek Hotel is Located in the magnificent park of Misurata City. Its 172 rooms, suites and apartments are all air-conditioned and equipped with private bathroom telephone, colour TV with satellite, mini-fridge and automatic alarm clock. A restaurant offering international cuisine and coffee shop will also help meet your requirements. There is also a swimming pool, open air cafeteria, theatre and show-room. We are proud of our 250 seat conference centre for hosting seminars, conferences and banquets. We also have a business centre and parking.

Rabta Hotel
Gharian
Tel: + 218 - 41 - 631970 / 73
Fax: + 218 - 41 - 631972

Rabta Hotel is located in the centre of Gharian City not far from offices, banks, boutiques, and shops. Its 68 rooms and suites are all air-conditioned and equipped with private bathroom, telephone, colour T.V. with satellite channels, mini-fridge and an automatic alarm clock. There is also a specialized restaurant offering international cuisine and coffee shop.

THE SOUTHERN REGION

Ghat

Ghat is located in the southwest of Libya, in the valley of Tanzoft surrounded by the mountains of Missak Milt. It was the only permanent Touareg settlement in the Sahara, strategically located on the route between the Mediterranean Sea and sub-Saharan Africa. Ghat has mostly been an eminent link between the north and the south of the African continent and has played the role of a thriving trading post and a resting place in the middle of the desert for the many caravans.

South of the Ubari road, hidden in rocky canyons, is one of Libya's best-kept secrets: the prehistoric rock carvings of the Black Plateau, and the Messak Stettafet of the Touaregs. They mainly feature animals which are no longer found north of the Sahara. Thousands of years ago, in the heart of what is now the Sahara, there was a civilization which reached a high degree of artistic perfection. In the Messak Stettafet the vast majority of carvings would seem to be over 4,000 years old.

WHERE TO STAY

El Jabal Hotel
Sebha
Tel: + 218 - 71 - 629407 / 9
Fax: + 218 - 71 - 629481

El Jabal Hotel is located in the magnificent desert of Sebha. Its 25 rooms/suites are all air-conditioned and equipped with private bathroom telephone, colour T.V. with satellite, mini-fridge and automatic alarm clock. A specialized restaurant offering international cuisine and coffee shop will also help meet your requirements. There is also a swimming-pool and open air cafeteria.

THE LIBYAN GASTRONOMY

Libyan cuisine is a mixture of Arabic and Mediterranean, with a strong Italian influence. Italy's legacy can be seen in the popularity of pasta on its menus, particularly macaroni. A famous local dish is couscous, which is a boiled cereal (traditionally millet, now often wheat) used as a base for meat and potatoes. The meat is usually mutton, but chicken is also served occasionally. Sherba is a highly spicy Libyan soup. Bazin, a local specialty is a hard paste, made from barley, salt and water. Dates, oranges, apricots, figs, olives and other fruits and vegetables are all readily available. For the most part, restaurants and cafes are used by foreigners.

International cuisine is available in some restaurants and in the larger hotels. Cafe opening hours outside the capital are somewhat limited, and most eating-houses close by 9pm. There are no nightclubs and bars. There are several cinemas in major towns, some showing foreign films. There are no theatres or concert halls. All alcoholic drinks are banned in Libya, in accordance with the laws of Islam. Bottled mineral water is widely consumed, as are various soft drinks. Fruit juices, particularly orange, can also be bought in shops. Local tap water is not recommended, as it has a slightly salty taste.

Libyan tea is a thick beverage served in a small glass, brewed with mint and often accompanied by peanuts. Regular American/British coffee is, of course, also widely available. Souqs in the main towns are the workplaces of many weavers, coppersmiths, goldsmiths, silversmiths and leatherworkers. There are numerous other stalls selling a variety of items including spices, metal engravings and various pieces of jewellery. There are large numbers of general stores in the cities, which sell everything from food to clothes.

W.T.S-1
Tripoli / Leptis Magna / Nalut / Ghadames / Gharian / Sabrata
8 Days / 7 Nights

Day	Itinerary	Detailed program
(01)	Tripoli airport	* Arrival at Tripoli airport transfer to the hotel. * Dinner & night at the hotel.
(02)	Tripoli	* A full day sightseeing tour of Tripoli including the old city, the Arch of Marcus Aurelius and the excellent museum. * Lunch at a local restaurant. * Dinner & night at the hotel.
(03)	Tripoli / Nalut /Ghadames 650KM	* Morning departure to Ghadames. * Stop in Nalut for a short tour of Gaser and the old city. * Lunch at a local restaurant. * Dinner & night at the hotel in Ghadames.
(04)	Ghadames	* A full day sightseeing tour of Ghadames including the old city & the house museum. * Lunch at traditional Ghadames house. * Short trip to the sand dunes. * Dinner & night at the hotel.
(05)	Ghadames / Gharian 550 KM	* Morning departure to Gharian. * Stop in places of interest on the way. * Lunch en route. * Dinner & night at the hotel.
(06)	Gharian / Leptis Magna / Tripoli 250KM	* Morning departure to the Roman city of Leptis Magna, one of the most impressive archaeological sites in the world including a visit to the excellent museum. * Lunch near the site. * Dinner & night at the hotel.
(07)	Tripoli / Sabratha / Tripoli	* Morning departure to the extensive ruins of Sabratha. * Lunch near the site. * Dinner & night at the hotel.
(08)	Tripoli hotel / Tripoli airport	* Transfer to the airport. * End of service

W.T.S-2
ARCHEOLOGICAL TOUR
Tripoli / Leptis Magna / Misurata/ Benghazi / Cyrene /ApOLlonia / Gaser / Libya / Sabrata
8 Days / 7 Nights

Day	Itinerary	Detailed program
(01)	Tripoli / airport	* Arrival at Tripoli airport transfer to the hotel. * Dinner & night at the hotel.
(02)	Tripoli	* A full day sightseeing tour of Tripoli including the old city the Arch of Marcus Aurelius and the excellent museum. * Lunch at a local restaurant. * Dinner & night at the hotel.
(03)	Tripoli / Leptis Magna / Misurata 200KM	*Morning departure to the Roman city of Leptis Magna, one of the most impressive archaeological sites in the world including a visit to the excellent museum. * Lunch near the site. * Dinner & night at the hotel.
(04)	Misurata / Benghazi 850KM	* Early Morning departure to the city of Benghazi, 850 km * Lunch en route. * Dinner & night at the hotel.
(05)	Benghazi /Cyrene / Apollonia /Gaser Libya /Benghazi 450KM	*Morning departure to visit the Greek cities of Cyrene & Apollonia and stop to view Gaser, Libya which houses some of the world best mosaics. * Lunch near the site. * Dinner & night at the hotel.
(06)	Benghazi / Misurata 850KM	*Morning departure to Misurata. * Lunch en route. * Dinner & night at the hotel.
(07)	Misurata /Sabratha/ Tripoli	* A visit to the extensive ruins of Sabratha. * Lunch near the site. * Dinner & night at the hotel.
(09)	Tripoli hotel / Tripoli airport	* Transfer to the airport. * End of service

Your
compass
to investment

www.winne.com

Online since 1996 and reporting about investment opportunities in emerging markets. The best database available on the net.

W.T.S-3
Tripoli / Nalut/Ghadames/New
Gaberoun / Germa/Ghat/Akakus
Alawinat/Alqarda/Gharian/Gaberoun
Lake/LeptisMagna /Sabrata
15 Days / 14 Nights

Day	Itinerary	Detailed program
(01)	Tripoli airport	* Arrival at Tripoli airport transfer to the hotel. * Dinner & night at the hotel.
(02)	Tripoli	* A full day sightseeing tour of Tripoli including the old city, the Arch of Marcus Aurelius and the excellent museum * Lunch at a local restaurant. * Dinner & night at the hotel.
(03)	Tripoli / Nalut /Ghadames 650KM	* Morning departure to Ghadames. * Stop in Nalut for short tour of Gaser and the old city. * Dinner & night at the hotel.
(04)	Ghadames	* A full day sightseeing of Ghadames including the old city & the house museum. * Lunch at traditional Ghadames house. * Short trip to the sand dunes. *Dinner & night at the hotel.
(05)	Ghadames / New Gaberoun 1000 KM	* Morning departure to New Gaberoun. * Lunch en route. * Dinner & night at the hotel.
(06)	New Gaberoun / Germa / Ghat 450KM	* Morning departure to Ghat. * Short stop to view the ruins of Germa. * Dinner & night at the hotel.
(07)	Ghat / Akakus / Alawinat 300KM(Off Road)	* Three full days exploring the cave paintings and rock engraving of Akakus area. * Camp for dinner & night.
(10)	Alawinat / Gaberoun Lake 350KM	* Departure to Gaberoun Lake. * Camp by the Lake for dinner & night.
(11)	Gaberoun Lake / Alqarda 140KM (Off Road)	* Full day exploring the surroundings Lakes of Mavo, Mandara, and Um Alma. * Camp for dinner & night.
(12)	Alqarda / Gharian 600KM	* Morning departure to Gharian. * Lunch en route. * Dinner & night at the hotel.

Day	Itinerary	Detailed program
(13)	Gharian /Leptis Magna / Tripoli 250KM	* Morning departure to the Roman city of Leptis Magna, one of the most impressive Archaeological sites in the world, including the excellent museums. * Lunch near the site. * Dinner & night at the hotel.
(14)	Tripoli /Sabratha / Tripoli	* Morning departure to the extensive ruins of Sabratha. * Lunch near the site. * Dinner & night at the hotel.
(15)	Tripoli hotel / Tripoli airport	* Transfer to the airport. * End of service

W.T.S-4
Tripoli /New Gaberuon /Germa / Ghat / Akakus / Alawinat /Gaberoun LaKe / Alqarda / Gharian / Sabrata
11 Days / 10 Nights

Day	Itinerary	Detailed program
(01)	Tripoli	* Arrival at Tripoli airport transfer to the hotel. * Dinner & night at the hotel.
(02)	Tripoli / New Gaberoun 900KM	* Morning departure to New Gaberoun. * Dinner & night at the hotel.
(03)	New Gaberoun / Germa / Ghat 450KM	* Morning departure to Ghat. * Short stop to view the ruins of Germa & the museum. * Dinner & night at the hotel.
(04)	Ghat / Akakus / Alawinat 300KM(Off Road)	* Three full days exploring the cave paintings and rock engraving of Akakus area. * Camp for dinner & night.
(07)	Alawinat / Gaberoun Lake 350KM	* Departure to Gaberoun Lake. * Camp by the lake for dinner & night.
(08)	Gaberoun Lake / Alqarda 140KM (Off Road)	* Full day exploring the surrounding lakes of Mavo, Mandara, and Um Alma. * Camp for dinner & night.
(09)	Alqarda / Gharian 600KM	* Morning departure to Gharian. * Lunch en route. * Dinner & night at the hotel.
(10)	Gharian / Sabratha /Tripoli	* Morning departure to the extensive ruins of Sabratha. * Lunch near the site. * Dinner & night at the hotel.
(11)	Tripoli hotel / Tripoli airport	* Transfer to the airport. * End of service

W.T.S-5
Tripoli / Gaberuon Lake /Mathendosh Gharian / Leptis Magna /Sabrata
8 Days / 7 Nights

Day	Itinerary	Detailed program
(01)	Tripoli	* Arrival at Tripoli airport transfer to the hotel. * Dinner & night at the hotel.
(02)	Tripoli / New Gaberoun	* Morning departure to new Gaberoun * Lunch en route. * Dinner & night at our camp.
(03)	New Gaberoun / Gaberoun Lake/ New Gaberoun	* Morning departure to view the Lake of Gaberoun. * Lunch by the Lake. * Dinner & night at our camp.
(04)	New Gaberoun /Methendosh/ New Gaberoun	* Departure to view the site of Methendosh. * Dinner & night at our camp.
(05)	New Gaberoun / Gharian	* Early departure to Gharian. * Lunch en route. * Dinner & night at the hotel.
(06)	Gharian / Leptis Magna / Tripoli	* Full day exploring the Roman city of Leptis Magna including a visit to the excellent museum. * Lunch near the site. * Dinner & night at the hotel.
(07)	Tripoli /Sabratha / Tripoli	* Morning departure to the extensive ruins of. Sabratha * Lunch near the site. * Dinner & night at the hotel.
(08)	Tripoli hotel / Tripoli airport	* Transfer to the airport. * End of service

193

TOUR OPERATORS

Al Madar Tourism Investments Services

Tel: + 218 - 21 - 4441953
Fax: + 218 - 21 - 3337429
E-mail: info@ Al.madargroup.com
Website: www.al-madargroup.com

Alfaw Travel

Tel: + 218 - 21 - 3340770
Fax: + 218 - 21 - 4802881
E-mail: Alfawtravel@yahoo.com
Website: www.Alfawravel.com

Alhassi Co for Travel and Tourism

Tel: + 218 - 21 - 4773461
Fax: + 218 - 21 - 4773461
E-mail: alhassicotour@yahoo.com
Website: www.alhassi.com

Alrayat-Tourism Services Co

Tel: + 218 - 21 - 4440632
Fax: + 218 - 21 - 4440632
E-mail: alraiat-tour@hotmail.com
Website: www.alrrayat.tk

Apollonian Tours

Tel: + 218 - 61 - 9097829
Fax: + 218 - 61 - 9097800
E-mail: apollonia- tours@hotmail.com
Website: www.apollonia- tours.com

Arwiqa Travel & Tourism

Tel: + 218 - 21 - 4910061
Fax: + 218 - 21 - 4910061
E-mail: arwiqa@hotmail.com

Asaria-Travel & Tours

Tel: + 218 - 23 - 620661
Fax: + 218 - 23 - 625507
E-mail: info@asariatours.com
Website: www.asariatours.com

Faradis for Travel

Tel: + 218 - 21 - 3340145
Fax: + 218 - 21 - 3332967
E-mail: info@faradistours.com
Website: www.faradistours.com

Farwa Tourism Company

Tel: + 218 - 21 - 4444869
Fax: + 218 - 21 - 4447305
E-mail: farwaco@hotmail.com

Fezzan Tours

Tel: + 218 - 71 - 634014
Fax: + 218 - 71 - 634014
E-mail: info@fezzantours.com
Website: www.fezzantours.com

RECOMMENDED

Ghadames Travel & Tourism

Tel: +218 - 21 - 4778224
Fax: +218 - 21 - 4778225
E-mail: ghadamestours@hotmail.com
Website: www.ghadamestours.com

Libyan Travel & Tourism Co.

Tel: + 218 - 21 - 3503282
Fax: + 218 - 21 - 4443455
E-mail: info@libyatravels.com
Website: www.libyatravels.com

GLOBAL EXPOSURE MARKETING.
TARGETING EMERGING MARKETS.

Global Exposure Marketing is an experienced, international marketing and distribution company, whose expertise is in emerging markets.

Our comprehensive range of services includes:

- The development and preparation of strategic marketing plans and programs.
- Key market segmentation and identification plans.
- The design and implementation of advertising and public relations strategies.
- Targeted product distribution.
- Industry trend analysis.
- Training in sales and marketing.

We are happy to work on small pilot projects or any other one-off assignments. However, we believe that the key to long term success lies in building robust long term relationships, between ourselves and our clients – and between our clients and their customers.

At Global Exposure Marketing, we aim to become the leading provider of marketing services which target emerging markets. Working alone, or alongside your mainstream agency, we provide the specialised knowledge, strategic advice and marketing expertise you need to capture a share of this expanding and rapidly developing market.

We centre our business on clients, rather than focus solely on the sales processes. By building long-term relationships with clients, not single-transaction deals with customers, we make them understand the value of the relationship.

We also offer consultancy, joint ventures and creative marketing support to businessmen with a desire to access African countries. For further details, contact our Business Development Manager on + 44 (0) 207 788 7825.

To find more information please visit www.globalexposure.co.uk

Ligh Way Service Company

Tel: + 218 - 21 - 3613155
Fax: + 218 - 21 - 3615910
E-mail: infi@hsc-libya.com
Website: www.hsc-libya.com

Rose de Tourisme Services

Tel: + 218 - 21 - 3335687
Fax: + 218 - 21 - 4449372
E-mail: rose.r@mailcity.com
Website: www.roselibya.com

Sh Awinat Tourism

Tel: + 218 - 71 - 635605
Fax: + 218 - 71 - 635605
E-mail: sh.awinat@yahoo.com
Website: www.awinat.com

Sukra Travel & Tourism

Tel: + 218 - 21 - 3340604
Fax: + 218 - 21 - 3340606
E-mail: sukratravel@hotmail.com
Website: www.sukra-travel.com

The Treasures Tours for Tourism Services

Tel: + 218 - 21 - 3334343
Fax: + 218 - 21 - 3339486
E-mail: Ltt-libya@mailcity.com
Website: www.ltt-libya.com

Touam Travel &Tours

Tel: + 218 - 21 - 3351510
Fax: + 218 - 21 - 3351509
E-mail: touamtraveltourism@hotmail.com

United Tours

Tel: + 218 - 21 - 4863959
Fax: + 218 - 21 - 4863959
E-mail: unitedtours@lttnet.net
Website: unitedtours-libya.com

North Africa Travel

Tel: + 218 - 21 - 3615894
Fax: + 218 - 21 - 3615894
E-mail: rort_africa_travel@hotmail.com

Tourism Services through Mediterranean Co

Tel: + 218 - 51 - 614460
Fax: + 218 - 51 - 614919
E-mail: meletan74@yahoo.com

Jannat Tours

Tel: + 218 - 51 - 624585
Fax: + 218 - 51 - 624586
E-mail: info@jannattours.com
Website: www.jannattours.com

International Centre for Exhibitions and Conferences

Tel: + 218 - 21 - 4442040
Fax: + 218 - 21 - 3332164
E-mail: info@libyatrade.com
Website: www.libyatradex.com

GADDAFI INTERNATIONAL FOUNDATION FOR CHARITY ASSOCIATIONS

The Gaddafi International Foundation for Charity Associations encompasses the following civil, non-governmental associations:

* Libyan National Society for Drug Control

* Human Rights Society

* De-Mining Society

* Southern Brothers Society

* Underprivileged Society

The Foundation is a non-governmental, non-profit organisation, which is based on the principle of voluntary work for the welfare of society and human development.

The Foundation strives to support civil participation through promoting and helping the social welfare societies and organisations to assume their proper role in the development of mankind and protection of his human rights.

The Foundation is an international organization: it's activities are not confined to a certain state or region.

GOALS AND OBJECTIVES OF THE FOUNDATION

To support the efforts of the associated societies and coordinate their activities.

To support the principle of volunteerism for the welfare of society and safeguarding of human rights.

To support the establishment of new NGO's based on volunteerism that will implement projects and activities to the benefit society.

To support vulnerable segments of society, such as children, orphans, the elderly, handicapped and deprived.

To work with welfare societies that develops and implement concepts to eliminate poverty.

To create awareness of social and humanitarian work and to perform relevant public information activities both on national and international level.

To strengthen and develop cooperation and interaction with international and regional organisations that have similar goals and objectives.

For further information please contact us at: Tel. (+218) 21 3351370/ 3336000   Fax (+218) 21 3351373/ 3331509

PO Box: 1101   Tripoli, Libya

www.gaddaficharity.org

VOCABULARY

Salamo alekom = Hello

Sabaa al kheir = Good morning

Masaa al kheir = Good afternoon

Tesbah ala kheir = Good night

Massalama = Good bye

Aiwa = Yes

Laa = No

Keif-halek = How are you?

Kwaiss, El Hamdollelah = Fine, Thanks

Khalas = Finish, done, ready

Hasma = Listen to me

Chai = Tea

Ahwa = Coffe

Shokran = Thank you

Suq = Market

Kwais = Good

Enta wen? = Where are you?

Menfadlek = Please

Yalah= Let's go!

COMPANY INDEX

Energy

Finance

Telecommunication

Leisure

Industry & Trade

Tourism

BIBLIOGRAPHY

The General People's Committe for Tourism

The information department of the Great Jamahiriya (www.nidaly.org)

Doing Business & Investment in Libya – Mohamed Ghattour & Co.

Responsibilities of a Branch Manager – Mohamed Ghattour & Co.

Foreing Nation Working in Libya – Mohamed Ghattour & Co.

www.libyaninvestment.com

www.libyaonline.com

The Academy of Graduate Studies, Tripoli

Become an eBizguides Exclusive Distribution Agent for your country!

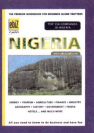
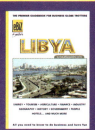
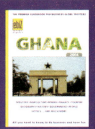

We are looking for local distributors in Africa
Join a successful project & team around the world.

Spain Representatives contact info: Tel +34 91 787 38 70 Fax +34 91 787 38 89 - info@ebizguides.com www.ebizguides.com

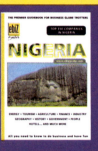
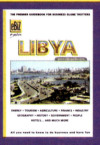
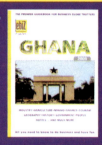

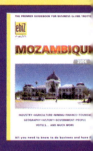

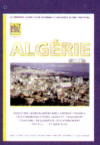

BOOK ORDER
(You can also order on-line at www.ebizguides.com)

- NAME:

- COMPANY NAME:

- ADDRESS:

- COUNTRY:

- PHONE (DAY TIME):

- CELL PHONE:

- FAX:

- E-MAIL:

NUMBER OF COPIES:

UNIT COST: 24€

TOTAL:

* A TRANSPORT CHARGE WILL BE ALSO BILLED: 15€

- QUANTITY:

- DATE:

- PAYMENT METHOD:

- SEND US A CHEQUE PAYABLE TO:

**PLEASE, FILL ALL THE INFORMATION ABOVE AND SEND IT TOGETHER
WITH CHEQUE FOR THE TOTAL QUANTITY TO:**

**MTH MULTIMEDIA.
Pº DE LA HABANA 169, ESCALERA B, 301
MADRID, SPAIN
TEL: + 34 91 787 38 70
FAX: +34 91 787 38 79
WWW.EBIZGUIDES.COM**